STEM PROs

STEM PROs

A Collection of Advice from Professionals in STEM

KEVIN J DEBRUIN

Originally self-published by DeBruin Enterprises, LLC
printed in hardcover by Amazon KDP 2024.

Self-published simultaneously by DeBruin Enterprises, LLC in eBook form.

IBSN Hardback: 979-8-9876402-3-4
IBSN eBook: 979-8-9876402-4-1
Library of Congress Control Number: 2024920863

Published 2024 in St. Petersburg, FL.
Printed in the United States of America.

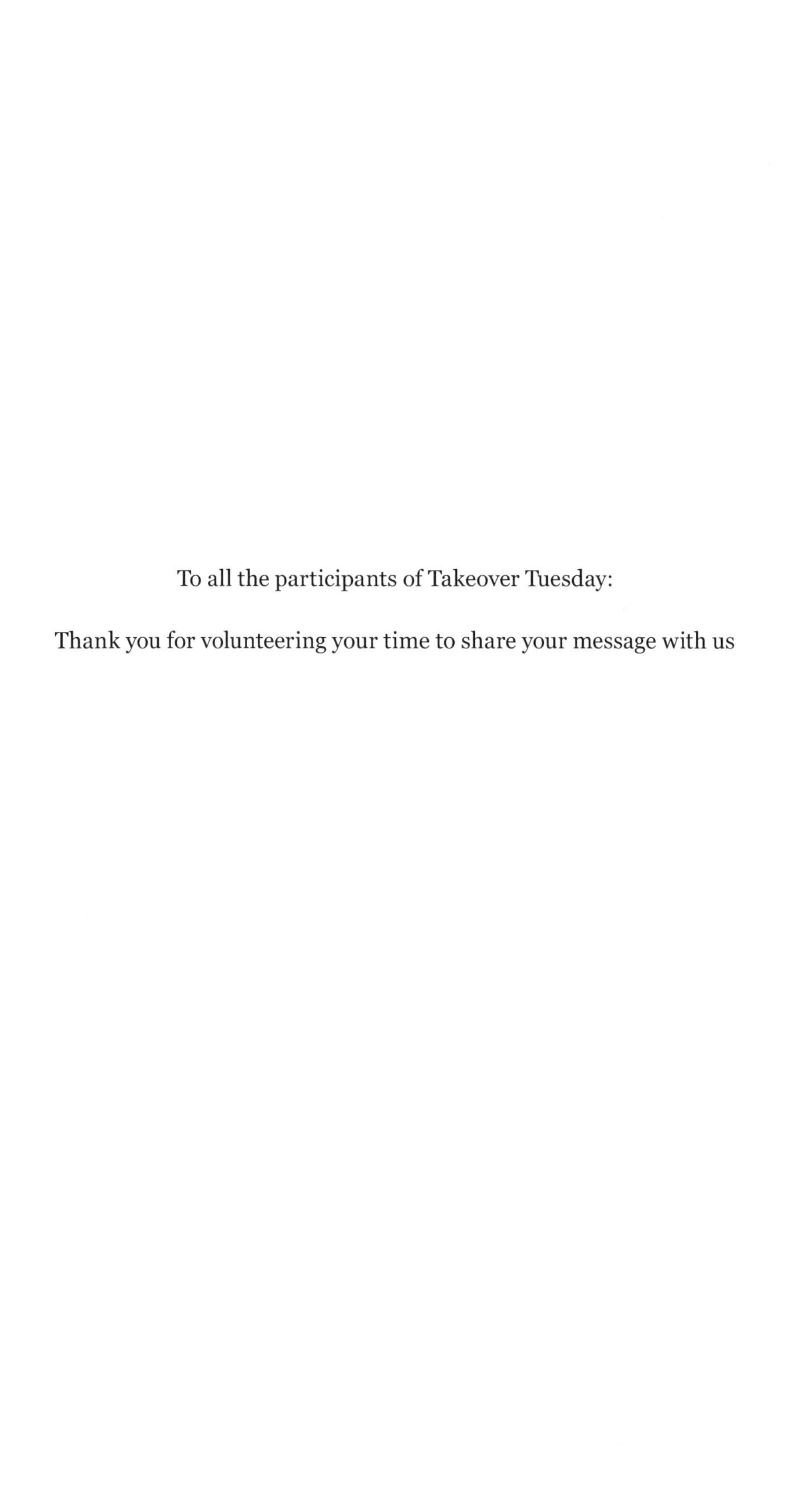

To all the participants of Takeover Tuesday:

Thank you for volunteering your time to share your message with us

CONTENTS

INTRODUCTION

When you hear STEM what do you think of? Is it the dated idea of a white man in a lab coat, glasses, and has a pocket protector? Maybe a scientist in a lab with a microscope or someone sitting behind a computer writing code? Or maybe a little more new age with a brightly colored created hairstyle hiking up the side of a volcano to sample lava?

STEM is an acronym coined in 2001 by the U.S. National Science Foundation (NSF) which stands for Science Technology Engineering and Mathematics. According to the Bureau of Labor Statistics, STEM jobs are projected to grow 10.8% between 2022 and 2032, compared to 2.3% in non-STEM jobs. That is almost 4x faster than non-STEM jobs!

Let's break three stereotypes of STEM with this book. One being that it is for men, that women do not belong. Vintage thoughts that were wrong in the first place. That was part of the thought-process behind *Takeover Tuesday* which you'll learn about in the next section. Two, that all STEM careers live in lab coats or in front of a computer; that is just a part of what some of the amazing careerists in this book do. From Antarctic expeditions to flying in fighter jets and outer space, you'll join researchers in Earth's most extreme environments chasing the newest scientific discovery highlighting some professions that ditch the lab coat for adventure. And three, that science and research are all consuming of a STEM career. The individuals in this book will shine light on the other aspects of their lives from making YouTube videos and creating art to competing in beauty pageants and fitness contests.

Takeover Tuesday

A DEI EFFORT | AMPLIFYING VOICES IN STEM

Takeover Tuesday was created in November 2020 by myself, Kevin J DeBruin. I wanted to find a way where I could use my unique set of skills and abilities to amplify underrepresented voices in STEM and bring awareness to amazing individuals and opportunities throughout the world. I also wanted to showcase professions outside of my role in the space industry. I desired to use any advantage I have with my privilege as a straight white male in the U.S.A. and give my platform to a female or BIPOC individual to spread their message and showcase themselves to my community. With this in mind, Takeover Tuesday was born.

Takeover Tuesday existed on social media, mainly Instagram, where I would give my login information to someone. They would log into my account and take control of the stories for a day (all day Tuesday) to share their message and experience with my community. There would also be a main feed post as well with the STEM individual's headshot and bio announcing the takeover participant. This occurred almost every single Tuesday for over three years with over 100 participants from around the world. Most of the participants were located in the United States, however there were several from Europe and South America that came onboard as well.

These stories ranged from still photos with text overlays, to videos, to Live Q&As, tours of labs and facilities (such as an animal sanctuary and an Antarctic expedition!). Participants volunteered their time and energy to showcase who they are and what their work was on my platform. I am extremely grateful for all those who agreed to participate and spread the good word of science to my community. I learned a ton myself!

Each person in this book I can confidently say that I will personally vouch for. It took a lot of trust to give full access to my social media to someone else. Some called me crazy and said it wasn't worth the risk because of the brand and community I built for myself. It didn't deter me at all. I know most of these people first-hand and others come from recommendations of friends. Yes there were some hiccups and close calls along the way as I vetted potential participants, but it all worked out! I put my trust in the scientific community, as I always do, and it didn't let me down. I applaud each of them for volunteering a day of their time on top of their careers, life, and hobbies to participate, and even more time to give thoughtful answers to each of the questions asked for this book. It's a chaotic dance at times to figure out availability for scheduling takeovers and to participate in an interview series to be collected into book format, and each of these individuals willingly danced with me to figure it out!

The Origin of this Book

I wanted to create a more permanent version of *Takeover Tuesday* available to everyone, on and off social media. I had read *Tribe of Mentors* and *Tools of Titans* by Tim Ferris and it gave me the idea for this collection. *Tribe of Mentors* asked the same 11 questions to hundreds of individuals from Presidents to pro athletes to entrepreneurs to actors to business owners. Ferris collected the responses into the book to offer "short life advice from the best in the world". I routinely go back to that book and flip around to different individuals depending on where I am at in life and business. It's been a terrific tool! Inspired by this resource, I set off to craft my own unique set of questions to ask all of my *Takeover Tuesday* participants. Most agreed and their answers are what follow. Some declined due to personal reasons, conflicts with employers, or exclusivity agreements.

What is STEM? STE(A)M?

STEM is the abbreviation for Science, Technology, Engineering, and Mathematics. In recent years there has been efforts to evolve STEM into STEAM where the A stands for Arts (incorporating both art and design). You'll notice the cover of this book has the four large quadrants to cover each area of STEM and then a smaller insert as a nod to A. Individuals in each area of STEAM are featured as STEM PROs and several crosscut multiple categories themselves.

The addition of the A for Arts in STEM to become STEAM is to include the art & design that is involved in the STEM fields. STEM was first expanded to STEAM in 2012 by the United States National Research Council. The Arts encompasses everything but not limited to design, communication, and creative planning. It's something that has always been present in successful and sustainable STEM projects over the years and is now being brought to light to showcase what else is involved in these areas. Jackie Speake, Ed.D., STEM education consultant and author of Designing Meaningful STEM Lessons says "STEM lessons naturally involve art (for example, product design), language arts (communication), and social studies and history (setting the context for engineering challenges)." Liz Heinecke, author of several STEAM project books for children, said ""Art and design have played a prominent role in STEM, whether it's pointed out to people or not." Now it's about time we shined a light on it!

A lot of STEM careers are not siloed in just one area of STEM. There are a lot of overlapping and crosscutting skills and areas of science, technology, engineering, and math. The examples that follow in this section display just some of the possibilities that exist in each area.

Science

The Sciences encompass all areas of Physics, Chemistry, Geology, Biology, and Medicine.

sci·ence
/'sīəns/
noun: **science**

1. the systematic study of the structure and behavior of the physical and natural world through observation, experimentation, and the testing of theories against the evidence obtained.

Middle English (denoting knowledge): from Old French, from Latin scientia, from scire 'know'. The main modern sense developed in the late 18th century.

Technology

Technology can be computer science, information technology, data science, and more.

tech·nol·o·gy
/tek'näləjē/
noun: technology; plural noun: technologies

the application of scientific knowledge for practical purposes, especially in industry.

"advances in computer technology"

- machinery and equipment developed from the application of scientific knowledge.
- "it will reduce the industry's ability to spend money on new technology"
- the branch of knowledge dealing with engineering or applied sciences.

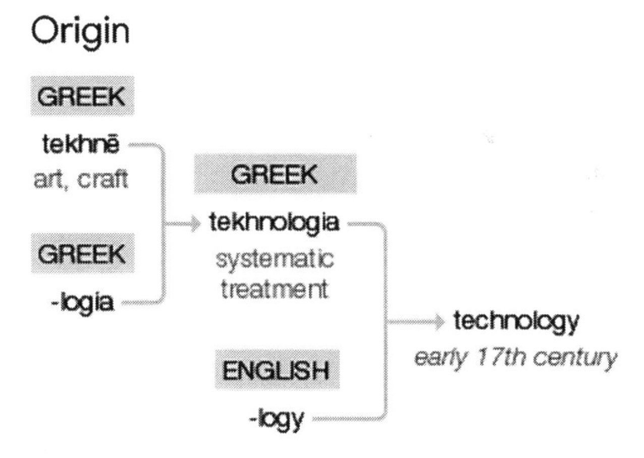

early 17th century: from Greek tekhnologia 'systematic treatment', from tekhnē 'art, craft' + -logia (see -logy).

Engineering

Engineering is the creation of something that did not previously exist. Some such areas are but not limited to: aerospace, civil, environmental, mechanical, electrical, software, computer, and nuclear.

en·gi·neer·ing
/ˌenjəˈniriNG/
noun: engineering

the branch of science and technology concerned with the design, building, and use of engines, machines, and structures.

- the work done by, or the occupation of, an engineer.
- the action of working artfully to bring something about.

Origin: Middle English (denoting a designer and constructor of fortifications and weapons; formerly also as ingineer): in early use from Old French engigneor, from medieval Latin ingeniator, from ingeniare 'contrive, devise', from Latin ingenium (see engine); in later use from French ingénieur or Italian ingegnere, also based on Latin ingenium, with the ending influenced by -eer.

Arts

Arts relates to all aspects of design, communication, and creative planning.

art
/ärt/
Noun: art; plural noun: arts

1. the expression or application of human creative skill and imagination, typically in a visual form such as painting or sculpture, producing works to be appreciated primarily for their beauty or emotional power.

Middle English: via Old French from Latin ars, art-.

Mathematics

Mathematics generally refers to statistics, data science, and applied mathematics which is using mathematical methods for application of physics, engineering, etc.

math·e·mat·ics
/ˌmaTH(ə)'madiks/
noun: mathematics; noun: applied mathematics; noun: pure mathematics

the abstract science of number, quantity, and space. Mathematics may be studied in its own right (pure mathematics), or as it is applied to other disciplines such as physics and engineering (applied mathematics).

- the mathematical aspects of something.

Origin: mid 16th century: plural of obsolete mathematic 'mathematics', from Old French mathematique, from Latin (ars) mathematica 'mathematical (art)', from Greek mathēmatikē (epistēmē), from the base of manthanein 'learn'.

STEM PRO's QUESTIONS

I chose the following 12 questions to ask each individual. Below is my rationale for crafting and keeping each question. I initially started with around 20 questions and downselected to the set here.

Why do you do what you do? / Do you have a defining/ah-hah/ eureka moment where you knew what you wanted to do?

What is something you wish you knew while in school?

What is something you wish you did differently when you first started working?

If someone wanted to do what you do, what's the best piece of advice you'd share with them?

If you could definitively answer one unanswered science question, what would it be and why? / What unanswered scientific question keeps you up at night?

What scientific discovery or event in history would you like to go back in time and witness?

What's a misconception about your line of work?

What's the best part of doing what you do?

What is a necessary evil in your industry?

Have you ever changed your line of work? If so, why and what was the change?

What purchase under $100 has improved your life: career or personal?

How has social media benefited or hindered your career?

It was quite tough to bring down the question list to these 12. I mean I would have loved to ask all the questions I had to everyone, but let's be realistic here. I'm asking people to volunteer their time and information to participate in this project. So I settled on the 12, just one more question than Tim Ferris asked in *Tribe of Mentors*. I figured he did enough homework to know that wasn't too little nor too much. So here is why I decided to choose each of the 12 STEM PROs Questions.

Why do you do what you do? / Do you have a defining/ah-hah/ eureka moment where you knew what you wanted to do?

The WHY, this is what drives people to do what they do and to persevere through adversity they encounter. I love hearing people's origin stories about why they do what they do. People also really enjoy sharing this because it's a core memory for them, the reason they are who they are today. It gives us a good look into who the person is and where they came from.

What is something you wish you knew while in school?

Always a good one. Basically a staple because school is when we are learning. Most people answer this in addressing the college season of education, where we are first really finding ourselves and being on our own navigating an educational world we're told we have to go into if we want a good job. Is that true? Let the STEM Pro's tell you what they think.

What is something you wish you did differently when you first started working?

This question came from what I did in my early days working my first "big boy" job at NASA JPL. Each of the more senior colleagues I encountered that I felt I vibed with, I would ask "If you could go back to when you first started working here (or the job you first had), what do you wish you did?"

I got things like "leave your work phone at work", "don't put work email on your personal phone", "say 'no' more", "don't work on weekends, even in the beginning, because it'll set a precedent that's hard to escape." Hindsight is 20/20, so if you're still in school or new to your career, heed this advice with great importance!

If someone wanted to do what you do, what's the best piece of advice you'd share with them?

Every single one of the individuals in this book is a role model. They get looked up to and more than one person has asked them how to be more like them. Envy or desire the positions or work of any of these STEM PROs? This question is what they'll tell you to be your own version of them.

If you could definitively answer one unanswered science question, what would it be and why? / What unanswered scientific question keeps you up at night?

This one lists out some of the most important work going on in today's science world. ORRRR highlights where we need to start researching or putting more funding towards. What are those unknowns people are looking to answer? What do we NOT know yet in this world of information that we live in? These answers are what will be at the bleeding edge of industries in the decades to come.

What scientific discovery or event in history would you like to go back in time and witness?

I chose this one because it seemed fun! What are some cool things in history it would be neat to see? I believe this question will also give us a glimpse into what makes these individuals tick; in a why do they do what they do or what excites them or occupies their mind so much they would want to experience it? Maybe you'll learn about some discoveries that you didn't think were that big or important. Is it the discovery of fire? Electricity? Gravity? Several mention the 1969 NASA astronaut moon landing.

What's a misconception about your line of work?

Time to break stereotypes about STEM! Let's get the **REAL** answers from the professionals. STEM careers and individuals have evolved over the years. Previously it was thought that tattoos could prevent you from ever getting a good job, BUT in 2023 NASA published a bunch of photos showcasing employees with tattoos showing how abundant they were in the workplace. So what are things that people mistake for truth in these STEM industries and positions? Let the STEM PROs bust those myths!

What's the best part of doing what you do?

What's the good stuff?! And it's not always what you think or even directly related to their work. Here we are able to get an insight into some of the indirect benefits that make people feel fulfilled and satisfied with that they get to go.

What is a necessary evil in your industry?

It can't be all sunshine and rainbows. I mean I love what I do now, and enjoyed my work at NASA (for a while...), but there are always going to be some negatives, some unpleasantries. I didn't want to ask "what do you not like about your job" or "what's the worst part of what you do". Because that's definitely a personal preference, where here I wanted to get into the weeds, the behind the scenes, pull the curtain back, look at industries and positions. In science there is a lot of grant writing and competing for funding, so what are some of the things you'll have to do in these industries/ positions to be successful? That's what this question aims to address.

Have you ever changed your line of work? If so, why and what was the change?

We get asked what we want to be when we grow up as children, and we are supposed to have an accurate and correct answer of what we'll be forever?? I don't think so. We stopped getting asked that as we grow up, BUT so many people learn more along the way that they decided to shift their trajectory and change their path. I had my defining moment at 10 years old to work for NASA, but then after a few years at NASA I had a new defining moment and changed course. I quit NASA, yeah crazy! Some of the people in the book have done that too! Some went to different space companies and others like me ventured out on their own. With these answers you'll see it's okay to pivot or even pull a u-turn in life.

What purchase under $100 has improved your life: career or personal?

This one I basically stole from *Tribe of Mentors*. Thanks Tim! His original question was "What purchase of $100 or less has most positively impacted your life in the last six months (or in recent memory)?" I love this question because it gives you an action to take, something to acquire that has a review behind it from someone I respect and look up to. It's also a monetary limit that's not too crazy and these people are saying it's worth it.

How has social media benefited or hindered your career?

Most, if not all, of the individuals asked have a presence on social media. It's actually how I met almost all of these individuals. When I was hiring interns at NASA, I would look at their social media. Yes it's a thing, we do it. Others do it. My colleagues and I wanted to know if we wanted to spend copious amounts of time with someone. We were going to be spending around 40-hours a week together and sometimes sharing our office with them so they had a place to work. Social media gave us a look past the resume. I was on a social media panel at the Space Symposium a handful of years ago and we discussed these effects. To have one, or not to have one. Which one to have?? LinkedIn seemed to be the best option because it is the professional social media. You'll see answers here that are mostly positive, but also a catch-22 where some wish they did less. I will say that I had several individuals decline to participate in this book because their employers were not enjoying how much they were putting themselves out in the world outside of their fulltime job. Which was quite unfortunate because this is to help their potential future workforce!

How to Use this Book?

The STEM PROs are not listed in any particular grouping of categories. I debated trying to go in the order of S.T.E.A.M. with these individuals, but so many of them crosscut several areas that it made it impossible to do so. Therefore a freestyle mix of them ensued. All are amazing and have great insights. As you read through feel free to bounce around, dog ear pages, highlight things, reach out to one you really resonate with on socials and tell them how much you enjoyed their insights. You may not resonate with every person nor all of the answers. This book is a collection of individuals from all around the country and world as well as their unique perspectives with their experiences. Some of these will "hit" for you and others you'll be quick to put aside. That's okay! Find the ones you relate with and learn more about them. But also get more informed on those who you may not see eye-to-eye with. I'm always learning, even in areas I disagree with, to be informed and aware of what's out there.

THE STEM PROs

Kelly Knight *(page 91)*
PhD Student in Science Education Research, MS Forensic Science & Forensic Biology, BS Chemistry & Forensic Chemistry, Associate Professor and STEM Accelerator of Forensic Science

Jacqueline Means *(page 99)*
Founder of the Girls Empowerment STEM Initiative; College student double majoring in Management Information Systems and Marketing with a minor in Neuroscience

Allison Cusick *(page 103)*
Biologist, Oceanographer, Adventurer, Antarctic Specialist, PhD in Biological Oceanography (2024), MSc Marine Biology (2021), MAS Marine Biodiversity and Conservation (2017), BSc Biology (2006)

Yasmin Dickinson *(page 111)*
PhD Student in Cardiovascular Medicine, Science Communicator, MS Molecular Biology, BS Biomedical Sciences

Elio Morillo *(page 117)*
MEng Systems Engineering & Design, BS Mechanical Engineering, Blue Origin, formerly NASA JPL

Dr. Sergio Sandoval *(page 121)*
PhD Aerospace Engineering, Guidance Engineer at NASA

Emily Seto *(page 127)*
Manager of Planetary Protection, Contamination Control and Research Scientist at Honeybee Robotics, formerly NASA

Dr. Yasmine Daniels *(page 131)*
PhD in Analytical Chemistry, Certified Industrial Hygienist in the US Government, Adjunct Professor, Author, mentor, mom

Dr. Shannon Cofield *(page 137)*
Ph.D. Geological Oceanography, U.S. Navy Veteran

Morgan Denman *(page 145)*
Space-focused Graphic Designer & Illustrator, NASA HQ

STEM PROs ANSWERS

STEM PRO

MAYNARD OKEREKE

MAYNARD OKEREKE: Better known as the Hip Hop M.D., is an award winning Science Communicator who helps encourage more minority & youth involvement in STEM by bridging the gap between science and entertainment.

IG: *@hiphopscienceshow*
TT: *@hiphopscienceshow*
X: *@thehiphopmd*
FB: *Hip Hop Science Show*
YT: *@HipHopScienceShow*

MAYNARD OKEREKE, better known as the Hip Hop M.D., graduated from the University of Washington with a degree in Civil Engineering. He is an award winning Science Communicator, having received both the Asteroid Award for "Best Streaming Content" and the People of Change Award for his community outreach efforts. His passion for science and entertainment, along with his curiosity for new innovation has taken him through an incredible life journey.

Noticing a lack of minority involvement in the S.T.E.M. fields, he created Hip Hop Science with the goal of encouraging minorities and youth to pursue more advanced career paths. His background in engineering, acting, music, business, and credible work in STEM make him uniquely qualified to engage on a wide variety of topics from an entertaining perspective. This is highly reflected in his speaking engagements and daily social media posts which provide both humorous and informative SciComm content.

STEM PROs Questions

Why do you do what you do? / Do you have a defining/ah-hah/ eureka moment where you knew what you wanted to do?

My goal with the platform has always been to spark curiosity in STEM by bridging the gap between entertainment and science. When I originally started making videos it was mostly from an entertainment perspective and I quickly realized that people were learning and getting educated by them. Representation in STEM is also highly important to me, and I think once I realized I had a voice that was resonating I wanted to really focus on encouraging more minority and youth involvement in these fields to help inspire the next generation.

What is something you wish you knew while in school?

I wish I knew I could actually combine my creative side with my science side. I grew up with the stigma that you couldn't do both and felt I spent most of my college years trying to juggle two different sides. This also affected my career decisions and I ended up feeling forced into professional choices that didn't fully resonate with who I was.

What is something you wish you did differently when you first started working?

I wish I walked in with more confidence. I was kind of stuck in imposter mode for so long, that by the time I really found myself I had already lost the passion for what I was doing.

If someone wanted to do what you do, what's the best piece of advice you'd share with them?

Never give up. Success truly doesn't happen overnight. If you're passionate about what you do, pour yourself into it and invest in growing. It will eventually pay off.

If you could definitively answer one unanswered science question, what would it be and why? / What unanswered scientific question keeps you up at night?

How exactly did dinosaurs live? We have a lot of evidence from fossil records and also use current living animals to compare, but I always wish I could snap my fingers and instantly be in the Jurassic period to get a clear visual of what, how, and where they interacted on Earth.

What scientific discovery or event in history would you like to go back in time and witness?

The aforementioned age of the dinosaurs! I want to witness the Earth moments before the asteroid hit and how it truly impacted the planet.

What's a misconception about your line of work?

That it's easy. I know social media glamorizes "influencer" lifestyle, but there is a lot of planning and behind the scenes work that goes into this. From networking to script writing, to all the time and family sacrifices you have to make to elevate. It's taken years of day in day out work to get to this level.

What's the best part of doing what you do?

Seeing and hearing the impact you've made. I always love getting messages from people of how my work has influenced or inspired them, or even helped their kids and students. That is by far the most rewarding part.

What is a necessary evil in your industry?

Staying relevant. In the sense that you have to consistently be "on". Even as an extrovert, I have moments where I just want to be off the grid for some time, but the nature of doing stuff in social media is constantly having to be in the mix and putting things out there for the public. It truly is a full time job.

Have you ever changed your line of work? If so, why and what was the change?

After graduating college I worked professionally as a Civil Engineer. My biggest transition was living engineering and going full time into entertainment work which was a complete 180 degree transition. My next transition which wasn't as steep was doing the science communication work I do now. I was able to combine many elements from both lines of work to create this platform and stay true to all the things I love.

What purchase under $100 has improved your life: career or personal?

My very first acting class I took in Seattle. It set me on the trajectory of stepping out of my comfort zone and connected me with some incredibly creative people who completely changed my perspective on life and my future.

How has social media benefited or hindered your career?

Social media has most definitely benefited my career. It's given me a platform to be recognized and opened the doors for so many new opportunities. Mostly it's connected me to an amazing community of scientists and other peers that I never knew existed before. They've inspired me, motivated me, provided opportunities for me, and helped instill confidence in my capabilities.

STEM PRO

GREGORY VILLAR

GREGORY VILLAR: MS Aeronautical Engineering, BS Physics, EDL Systems Engineer, Blue Origin, formerly NASA JPL

IG: *@GregoryVillar*

GREGORY VILLAR was a Systems Engineer at NASA JPL. He has held several roles on the Mars Science Laboratory flight project that provided him experience in all phases of the mission. Gregory tested operations for launch, cruise and landing, and worked Mars time surface operations with the responsibility of planning rover activities daily. In addition, he tested software for the sampling system using Curiosity's twin test rover on Earth.

Gregory graduated with a bachelor's degree in Physics and master's degree in Astronautical Engineering. He participated in several internship programs at NASA's Jet Propulsion Laboratory, before being hired as a Systems Engineer in 2010. Gregory spent 8-years working on the Mars 2020 flight project, as part of the team that was responsible for safely landing the Perseverance rover on Mars. He is now emploat Blue Origin.

STEM PROs Questions

Why do you do what you do? Do you have a defining aha Eureka moment where you knew what you wanted to do?

Hmm. I'm guessing this is for STEM, so not what I'm doing today, but getting into NASA it was, you know, the typical thing where I grew up watching space movies, and I think the one I always go to was Independence Day with Will Smith. Basically when they uhm they take the alien spacecraft, and they go up into the atmosphere, and it's just a blanket of stars, and Will goes you don't understand I've been waiting for this my whole life.

What is something you wish you knew while in school?

How to do finance. So, I wish you know, going on this entrepreneur journey now, it's actually quite fascinating that there's a lot of things I wish that school taught us that wasn't STEM more like, you know, financial management, business related things.

What is something you wish you did differently when you first started working?

Hmm, I think this one would be being more social. When I started at JPL for two years as an intern in astrophysics, I only interacted with my mentor and maybe like the one or two other interns. And so far so, I didn't even know JPL did flight projects. That was the year that Phoenix landed and I didn't even know. And as you know, I got really social when I became an engineer and I think that really kind of catapulted my career when I built up a network. Over like five to ten years. So, I really wish I got more social in the very beginning and built up that network.

If someone wanted to do what you do, what's the best piece of advice you'd share with them?

Ah man, it's so cliche, but whether it's space or now entrepreneurship, it's kind of love what you do. It's so cliche, but I really think it helps because when I was at JPL, I never felt like I was going into work and I think that's kind of true now in this journey that I'm going through. Like there's stress and stuff, but at the end of the day I love what I do and so you know when you love something you'll be more motivated to do that work and continue on.

If you could definitively answer one unanswered science question, what would it be and why? What unanswered scientific question keeps you up at night?

Man, these are so deep. Alright, if you could definitively answer one unanswered science question, man, I mean I don't know if this counts but I really like the idea of time travel, I mean if we can solve that, I mean I know like in theory it's not possible to travel. Backwards into time, although we can travel forward, but you know just growing up like one of the sci-fi movies I enjoyed was Back to the Future and so all the, and Terminator, like all the time travel type of movies, so I think that'd be cool to solve that, or "solve", you know, in quotes.

What scientific discovery or event in history would you like to go back in time and witness?

Wow. For some reason I thought about Oppenheimer here, but I wouldn't go to that. Would be really cool. What scientific discovery or event in history would you like to go back to in time and witness? Man, I'd say the moon landing, not to cop-out again. I mean, just to experience that in person would have been really cool. I would have to think of something else. Like, for some reason I thought about human discovery. Discovering fire. I think that would have been really cool too.

What's a misconception about your line of work?

As a systems engineer, and you get this, I feel like people think like we know a lot about equations and, I mean we do, but like you know the rocket scientists part of things that's not everyone at JPL or NASA right, like there's even like business people that contribute to it too. So I'd like to say that probably systems engineering being more of like a soft skill in a way, but something that's very critical for complex projects, so. Yeah, I don't think people know that I don't do like super complex equations and analysis, at least in systems engineering.

What's the best part of doing what you do?

That one's easy. At JPL, it was by far the people that I worked with on top of it. Everyone being brilliant, of course, everyone had awesome personalities, and especially the team I was on, like, we were basically family, and kind of coming into work was even more of a joy because I got to work with people that I loved being around.

What is a necessary evil in your industry?

Necessary evil? Well, at least in NASA. I think a necessary evil in our industry is budgets. Probably not so much so for like SpaceX and Blue Origin, but you know, NASA is limited by budget and it sucks because, you know, JPL just went through the layoffs. But I think in a way, budget and schedule really helps you define a problem and execute it. I think with an infinite budget and infinite schedule, we can just always make things better and so there's something to be said about being constrained by those two things.

Have you ever changed your line of work? If so, why and what was the change?

Well, yeah, I did that a few times. Graduating high school, I thought I was going to be a lawyer, an accountant/lawyer. Growing up, I always wanted to be an astronaut and for some reason in high school, just watching the debate team inspired me to be a lawyer. That was short-lived because I went to college. Did my first two years and realized my passion was in science, so I switched to physics. Did research in astrophysics for two years at JPL and realized I don't like data analysis. So then I switched to systems engineering and then did systems engineering for like 10 to 15 years, depending how you count that. Which I loved by the way. And then at some point just felt kind of complacent. And I like I wasn't growing anymore and I lost that passion and now I'm trying entrepreneurship. So I'm having the time of my life although making 10 times less, but still having a blast.

What purchase under $100 has improved your life,career, or personal?

Damn, that's tough. $100. Oh okay, I know. I never really read growing up, but in the past two years I've been reading a lot, and I'd say like all these books, you can get books from $10 to $20 so books for sure. I mean if I, $100, I'd go back now and be like hey look, read these books. A lot to choose from some I'll throw out. Obviously The Alchemist, Rich Dad Poor Dad, and Atomic Habits, I would say top three.

How has social media benefited or hindered your career?

Oof I think at NASA it didn't really help much, I didn't really become an influencer. And I think in life, I would say maybe it was like a net negative, and that's kind of why I'm off of social media now. I think it's a distraction if you're not, if you're not catering your, your social correctly and definitely the mental health aspect of it. So, you know, I've been off of social for about two years, and that has really helped not, not experience FOMO, and you know, along with things like meditating, reading, working out every day, and reading, meditating, working out, and what's the other thing I do every day? Read, meditate, work, oh journal, and journaling, so yeah, yep.

STEM PRO
FABER BURGOS

FABER BURGOS: Science Communicator, Guinness Record Holder, Modern Languages Professional, Science influencer.

FB: *Faber Burgos*
TT: *@faberburgos*
IG: *@faberburgos*
YT: *@FaberBurgosSarmiento*

FABER DAVID BURGOS SARMIENTO, the Colombian science enthusiast and influencer, made a name for himself by pushing the boundaries of exploration. In October 2020, he achieved recognition by sending a selfie to the stratosphere using a homemade weather balloon, capturing the sun and the moon simultaneously, setting the stage for a Guinness World Record in 2021 for the most-viewed video on Facebook Creators. But his journey didn't stop there. While working in recycling to support his family, Burgos discovered 400-million-year-old Ammonoid fossils during his youth, a testament to his unyielding passion for science.

In another groundbreaking mission, he became the first Colombian to reproduce a recording at high altitudes during a balloon launch, reaching 35 kilometers and documenting Colombia's Ritacuba peak.

In 2021, he won an InstaFest award and was named the most influential personality at the Latin Plug Awards. He even sent a handwritten letter to space via SpaceX's Falcon 9 rocket, which traveled over 217 million kilometers setting a new record on this kind of experiment.

October 2022 saw Burgos collaborating with Juan Valdez to send coffee to space for International Coffee Day and being recognized as one of Colombia's Ten Outstanding Youth.

Moreover, he won an international contest to experience microgravity aboard a reduced-gravity aircraft, becoming the first Colombian chosen for such an opportunity by Space for Humanity.

STEM PROs Questions

Why do you do what you do? / Do you have a defining/ah-hah/ eureka moment where you knew what you wanted to do?

I do science on social media to inspire kids. I'm pretty sure that we can build a different society using science as the biggest tool to transform our behaviors and our reality. Right now the world has many challenges the most big is Global warming to solve that we need a different human being based on science. I'm working on it even been a professional of another area. To get that different society I guess that is more easy to learn through experiences than just theory. For that reason I share experiments of any kind once a week on my networks to connect kids with science in a fun way. In these experiments many professionals are around and behind me so we build daily a community where science is the most important thing. Of course I had an eureka moment I found an ammonite fósil when I was a kid. This creature let me realize that this world was science, is science, and will be science.

What is something you wish you knew while in school?

I had two dreams when I was at school. The first one of course was to become an astronaut and the second one to become a paleontologist. I know that are too different between them but right now being an adult I'm sure that I can have on myself a little bit of both

What is something you wish you did differently when you first started working?

When I started to work it was fun because the TV shows and soap operas here in my country are very popular. In those days, I wanted to be an engineer and then become an astronaut but it was late (for budget) because I picked another career for my life. Sometimes on the life you have to take decisions that will be with you forever.

If someone wanted to do what you do, what's the best piece of advice you'd share with them?

Definitely I would recommend never stop studying. I'm a science communicator based on humanities and I learn everyday things about everything. This job gave me the chance to meet incredible people from all over the world (astronauts, engineers, scientists, technicians, etc) and I always get the same conclusion: these people get their dreams because they study to make them a reality. That's it. There is no secret. If you want to be a science communicator on social media you must be patient and persevere. It's a hard world but you will succeed if you never give up. My top secret is: make science something that everyone can understand.

If you could definitively answer one unanswered science question, what would it be and why? / What unanswered scientific question keeps you up at night?

I have two: First one: is there life after death? And the second one: is there life on other places of this universe? both right now are unanswered questions for science but I think that both has all the answers for many questions that are unanswered.

What scientific discovery or event in history would you like to go back in time and witness?

I would like to travel to the first moment of our universe and watch how everything was made. I know that was the Big Bang happening out there but I think that this universe could be older than we think.

What's a misconception about your line of work?

Some people think that I'm a scientist and that I could answer everything and of course not. I'm just a science enthusiast or a science communicator and maybe I can offer some answers but there are a lot of questions that I can't answer and I will never get those answers.

What's the best part of doing what you do?

The smiles of the kids that watch my work daily on social media that's priceless. It's cool when they send their school projects to my email and I realize that I'm doing a good job. Future engineers and steam professionals are coming.

What is a necessary evil in your industry?

Maybe the mental competence of other social media creators, that's horrible. But could be necessary to improve processes

Have you ever changed your line of work? If so, why and what was the change?

Yes, in my life I have to do a lot of things to get money. I was an actor, moving helper, and a recycler. And all of these activities were totally out of my career and what I wanted for my life but everything in this life is a process and takes time to get your dream job.

What purchase under $100 has improved your life: career or personal?

Some years ago I bought a weather balloon to make an experiment in Colombia. It cost just 90 bucks. That's changed my life forever since the day that I did this experiment using that balloon, nothing was the same.

How has social media benefited or hindered your career?

It gives me the chance to give voice to those that make science everyday and not have a voice. I really love to be a science amplifier and that's the power that social media has given me. Nothing in this life has a better value than that.

STEM PRO

ELIANA SHERIFF

ELLIE SHERIFF: Science & Technology Communicator, @EllieInSpace on YouTube, Former TV News Anchor/Reporter

IG: *@elianainspace*
TT: *@ellieinspace3*
X: *@esherifftv*
FB: *Ellie in Space*
YT: *Ellie in Space*

ELLIE SHERIFF is a former TV News anchor/reporter with almost a decade of experience interviewing people from all walks of life.

Ellie discovered Starlink in 2021 and felt a calling to cover this new technological advancement from SpaceX to serve rural and underserved communities around the globe. She quickly began to expand her coverage to focus on all things SpaceX including the rapid iteration of the Starship program and the push by Elon Musk and a team of some of the most incredible engineering minds of our time to get humanity on Mars.

Ellie quit her job in TV news in May of 2022 to go all in on her YouTube channel. She is now full-time producing content related to space exploration and futuristic technologies. Ellie brings a unique twist to the space scene using her journalism background to interview experts and beloved internet personalities. She is currently living in Austin, Texas and loves to work out and explore the comedy scene of Austin when she's not cranking out videos. Ellie also loves to travel.

STEM PROs Questions

What do you do? / Do you have a defining/ah-hah/eureka moment where you knew what you wanted to do?

At first I would have said I just make videos for fun in my spare time about what SpaceX and Starlink are about. But now that I've had my Ellie in Space channel for three years, my mission statement is clearer. I am taking my skills as a TV news reporter and trying to bridge the gap between the public and space enthusiasts about what's going on in the emerging commercial space sector. I knew I wanted to go full time when my passion for my channel outweighed my passion for reporting on TV.

What is something you wish you knew while in school?

I wish I knew that you don't have to force yourself to go to school just because your parents want you to. I feel like college is a great option, but I am not sure if I got as much out of it as I should have for the premium I paid. I wish I had waited until I had a better idea of what inspired me and what I was passionate about. I also wish I had been more aware of how short your time in college really is. I would have been more involved and tried more things within my campus if I had been aware of how fleeting the college experience really is.

What is something you wish you did differently when you first started working?

I am grateful for the path I took because it got me to where I am but it did seem to take me a long time to believe in myself to the extent I do today. I wish for sure I would have put myself out there and started social media way long ago. The digital footprint really does build over time and so many people, including myself for a while, are scared to even put themselves on the internet for fear of failing. Well, turns out I didn't fail and I wish I had started sooner.

If someone wanted to do what you do, what's the best piece of advice you'd share with them?

Don't give up and keep creating. Also, start uploading TODAY! It takes a while to hit your stride and even when you feel like you have, there are setbacks and bumps along the way. Some videos you think will soar end up bombing and others you had low expectations for end up really resonating. So, just keep going! And launch that channel/idea/passion!

If you could definitively answer one unanswered science question, what would it be and why? / What unanswered scientific question keeps you up at night?

I mean, of course, let's address the Elephant in the room. Or aliens, and lack thereof. I would love to know if there are other intelligent forms of life in the universe and when we will, if ever, make contact.

What scientific discovery or event in history would you like to go back in time and witness?

I would love, love, love to see the dinosaurs. Without a question, these mighty beasts lived for SO much longer than us, yet we hardly know anything about them. I want to know what they actually SOUNDED like, how they looked, and what it would feel like to be a fly on the wall in that time period.

What's a misconception about your line of work?

I think a misconception a lot of people have is that, well you do YouTube, but you don't really "work"... while I do have a flexible schedule, the notion that YouTubers or other creators don't really have a job is preposterous. On the contrary, it's almost nonstop work, but sometimes it's fairly easy, (such as posting to social media to grow my account and feed algorithm) but other times, I may have to grind really hard during a Starship launch and work nonstop to gather, edit, and publish content. So yes, it's not a 9-5, but it does have its challenges!

What's the best part of doing what you do?

The best part is the community building and unique opportunities. I feel like I've met so many people online, and then some eventually in person from my channel and I am so grateful for all the connections I've made!

And I have traveled to places and met people/done things I would have NEVER been able to without the recognition of the channel! Never did I ever think Elon Musk would subscribe to me on X, I'd watch the first Starship rocket launch with Maye and Kimbal Musk, that I would interview Scott Manley at his house, that I would win a Zero G contest and fly for free, that I would interview some of the most important people in the history of SpaceX and more.

What is a necessary evil in your industry?

Online trolls and dealing with mean comments. Because I am putting myself out there online, I will attract some keyboard warriors as my channel grows. It's unfortunate but true that the bigger the channel gets, the more attention it will receive and statistically speaking, not all comments will be nice, but that's ok!

Have you ever changed your line of work? If so, why and what was the change?

I worked in TV news rising the ranks from producer, to reporter, to anchor over almost a decade. Near the end of my time in news, was when I decided I know how to tell stories, I have the skills to make my own brand, and I need to go for it and pursue SpaceX coverage full time. I'm still not satisfied with where my channel is, but successful people always stay hungry! I am trying to be proud of where I've come from more though. I think what I did in TV news was very similar to what I currently do in YouTube, but the metrics are different. Before, I would consider making my deadline for my package and live shot a success, now it only feels successful if I have tons of views and more subscribers. You go from being paid a steady salary to having to make all of your money through your channel AND becoming self-employed, self health insured, etc. It can be scary!

What purchase under $100 has improved your life: career or personal?

I really wish this question was in the 300 dollar range because for sure my DJI dual mic set (NOT sponsored) has made a HUGE difference for me. I am able to use my iphone and have quality sound and long battery life so I can make content and record almost all day with no issues. But under 100 dollars, I'd have to say it is TSA Pre Check. It is good for 5 years and I travel a lot and it surely has saved me so much time traveling through airports. Plus, you can keep your shoes on!

How has social media benefited or hindered your career?

Social media essentially is my career so it's put me in a position for unlimited earnings potential and a more flexible life than TV news ever allotted. But to say I am not addicted to social media and have an even more unhealthy relationship with it now would be a lie. It definitely is hard to separate your self-worth and validation from the views and engagement sometimes, but overall the positives outweigh the negatives and I am truly grateful for what it's allowed me to do in life and the doors it will continue to open! I truly hope to be on the sidelines, interviewing the first people who will go to Mars when the time comes! Ad Astra!

STEM PRO
EDUARDO FLORES

EDUARDO FLORES: BS Mechanical Engineering, Creator @engineeringmemesguy, coFounder Lumis Automation

IG: *@engineeringmemesguy*

EDUARDO FLORES has a bachelors in mechanical engineering and graduated from Chico State in California. He is a creator, and founder of engineering memes guy IG page who receives millions of views a week. Host of Women in STEM Wednesday who are live every week. He is also one of the founders of Lumis Automation who focuses on integrating fully industrial automation solutions.

STEM PROs Questions

Why do you do what you do? Do you have a defining/ah-hah/ eureka moment where you knew what you wanted to do?

I've always been captivated by the idea of creating and designing, a passion that traces back to my childhood in Usmajac, Jalisco, Mexico, where I built miniature cities from dirt and imagined advanced technological features within them. This passion was solidified during my Mech 140 class at Chico State, an introduction to machine design, which set my path in the field of mechanical engineering.

What is something you wish you knew while in school?

I wish I had known that my journey through mechanical engineering would take six years instead of four. Discovering the need for additional time in my third year was surprising, but it taught me the importance of embracing my educational path at my own pace, relieving the pressure to conform to a standard timeline.

What is something you wish you did differently when you first started working?

When I first started working, I wish I had been more confident in sharing my ideas during brainstorming sessions, regardless of my experience level. I've learned that every perspective is valuable and that speaking up can contribute significantly to a team's success and innovation.

If someone wanted to do what you do, what's the best piece of advice you'd share with them?

I would advise anyone aspiring to a STEM career to seek out internships or mentorship opportunities. Engaging directly with the field can provide invaluable insights, confirm your interest and passion, and offer practical experience that is essential for career development.

If you could definitively answer one unanswered science question, what would it be and why?

The question of how to achieve rapid interplanetary travel fascinates me. Solving the mystery of bending space-time for teleportation could revolutionize our understanding of the universe and significantly advance human exploration and expansion beyond Earth.

What scientific discovery or event in history would you like to go back in time and witness?

I would choose to witness Isaac Newton's formulation of the principles of calculus. Observing the moment when Newton laid down the foundations of this critical mathematical field would offer profound insights into one of the most influential discoveries in science.

What's a misconception about your line of work?

A common misconception is that running businesses and creating content is straightforward and lucrative. In reality, these roles are multifaceted, requiring a blend of skills in engineering, management, sales, and marketing, and they demand significant effort without guaranteed financial success.

What's the best part of doing what you do?

I find great joy in the creativity of engineering, the success of completed projects, the global connection through content creation, and the rewarding process of selling ideas and generating leads through marketing. These aspects make my work fulfilling and impactful.

What is a necessary evil in your industry?

Enforcing company rules to maintain a culture of respect and inclusion is a necessary evil. While it's challenging, it's essential for the health of the company, requiring tough decisions like letting go of toxic individuals to foster a positive and productive work environment.

Have you ever changed your line of work? If so, why and what was the change?

I transitioned from being an automation engineer to running two companies and becoming a content creator. This shift allowed me to explore a broader range of opportunities, leveraging my engineering skills in new entrepreneurial and creative contexts since June 2018.

What purchase under $100 has most improved your life: career or personal?

Learning Photoshop for under $100 has been a game-changer, enhancing my ability to sketch ideas for sales proposals and create engaging content. This skill has provided a competitive edge in my proposals and has been crucial in generating significant revenue and supporting my content creation endeavors.

How has social media benefited or hindered your career?

Social media has brought networking opportunities, conference invitations, and some income, building a supportive community and promoting my businesses. However, it requires significant time investment without always proportional financial return. Despite this, its overall impact on my career has been positive, offering valuable connections and visibility.

STEM PRO

JANET ESCOBAR

JANET ESCOBAR: Author of *The Latina Trailblazer*, angel investor, Venture Capital Scout, Women in STEM Advocate, consultant for Fortune 500 companies, BS Business Analytics

IG: *@Janet.escobar88*
Website: *Janet-escobar.com*

JANET ESCOBAR is a consultant for Fortune 500 companies, currently working with Microsoft's defense and intelligence team to develop their AI thought leadership. Alongside her consultant role, she scouts for two venture capital firms investing in preseed and seed startups. Her focus is on deep technology sectors, such as AI, IoT, robotics, and 5G.

Outside her professional commitments, Janet dedicates her free time to advocacy work. She volunteers for Latinas in Tech, a non-profit organization with the mission to connect, support, and empower Latinas in the technology field. Additionally, she serves as a mentor for students through Braven, a program that equips first-generation college students, students of color, and those from low-income backgrounds with skills, networks, experiences, and confidence to secure strong first jobs.

STEM PROs Questions

Why do you do what you do? / Do you have a defining/ah-hah/ eureka moment where you knew what you wanted to do?

I'm in venture capital because I love surrounding myself with visionaries who are optimistic about the future. Entrepreneurs identify a societal challenge, believe they're the ones who can solve it, and dive into action with determination. Witnessing their drive is truly inspiring. It's an industry where imagination, creativity, and business strategy meet. My "aha!" moment happened during the first pitch competition I attended. Each founder took the stage, succinctly presenting their business models aimed at tackling real-world issues close to their hearts. Immersed in this environment, it felt like I had discovered my calling, and since then, I've never looked back.

What is something you wish you knew while in school?

Networking is more important than grades. Your network can open doors for you and even expedite your career progression. Networking can be intimidating, particularly for introverts like myself, but it doesn't have to be daunting. I view networking as simply cultivating friendships. Just as you support your friends in achieving their aspirations and connect them with opportunities, networking operates similarly. It involves asking probing questions to discern their objectives and leveraging your resources—both people and knowledge—to help them.

What is something you wish you did differently when you first started working?

I wish I would've adopted an ownership mindset earlier, instead of an order-taker mentality. This sentiment might resonate with many first-generation Americans: we're often taught to unquestioningly follow instructions, keep a low profile, and simply be grateful for our jobs. This is terrible advice. Initially, I struggled to take ownership of projects, merely seeking instructions to execute. Gradually, I grew in confidence and started suggesting solutions and recommendations, and now, as a consultant, I get hired to advise executives on what they should do to meet their goals.

If someone wanted to do what you do, what's the best piece of advice you'd share with them?

For my job as a venture scout and consultant, my advice would be the same: learn how to talk to people. Active listening, asking good questions, and reading the room are a few things you need to be great at. You'll be surprised how many people lack these skills. Charismatic people have a good blend of competence and confidence. Be knowledgeable in your field, but don't disregard learning soft skills.

If you could definitively answer one unanswered science question, what would it be and why?/ What unanswered scientific question keeps you up at night?

Extraterrestrial life. There are so many galaxies it's hard to believe we're the only intelligent civilization.

What's a misconception about your line of work?

A common misconception about venture capital is that it's exclusively reserved for the privileged few. In reality, there isn't a single defined path into this field; numerous avenues lead to involvement in the industry. However, it's crucial to recognize the significant disparity in funding allocation, with less than 2% of venture capital funding directed towards women, and an even smaller fraction towards women of color. Additionally, the representation of women in decision-making roles within venture capital firms remains alarmingly low. To foster a more equitable society, it's imperative to have diverse voices in positions where they can allocate resources to address the challenges faced by marginalized communities.

What's the best part of doing what you do?

I get to be a part of the future and learn about emerging technologies that are going to influence our society and future generations.

Have you ever changed your line of work? If so, why and what was the change?

Yes! I love the idea of trying new things and pivoting. Change is inevitable in entrepreneurship or if you're a multi-faceted individual. It's okay to try things and change your mind about them. That's what growing is all about!

I studied business analytics intending to become a data scientist, but after graduating, I craved a role where I could foster my creativity, so I went into marketing. I was a growth marketer for a fintech startup, then became a consultant for Fortune 500 companies. In my role as a consultant, I get to practice client relationships, project management, and content strategy. However, while I love working with the largest technology companies in the world, I love the fast-paced energy of startups, so I'm currently transitioning into venture capital full-time.

In between these career shifts, I also tried website design and considered being a full-time memoir writer.

What purchase under $100 has improved your life: career or personal?

Career—Books: Mastery by Robert Greene, How Successful People Grow by John C. Maxwell, and Managing Oneself by Peter F. Drucker.

Personal—the Bible, therapy, and EFT tapping

How has social media benefited or hindered your career?

Social media is what you make of it. I've leveraged it to network, leading to exciting opportunities and connections with inspiring individuals who share similar interests. I met Kevin through social media and now I'm a part of this incredible project. I think it's worth it to take the time to build your online presence because it often serves as the initial point of contact for others. Make sure your profile reflects who you are and what you stand for.

"Nothing in life is to be feared,
it is only to be understood."

—Marie Curie, physicist, chemist, and winner of two Nobel Prizes known for discovering radioactivity, radium, and polonium

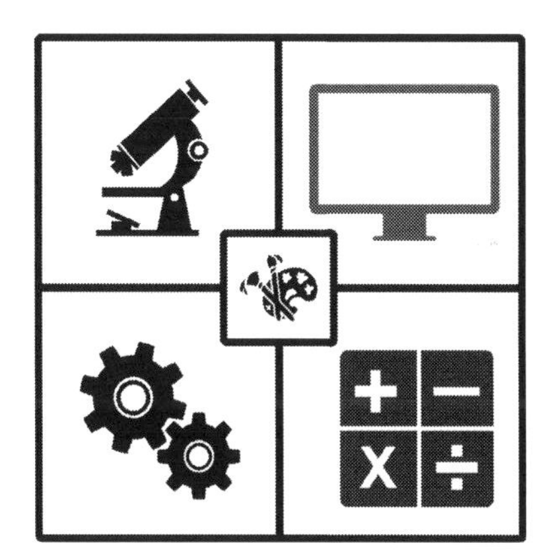

STEM PRO

OUMAIMA MOUTTAKY

OUMAIMA MOUTTAKY: Biologist & MB-engineering. Business Manager within the medical sector, specializing in Regenerative Medicine & Cancer immunotherapy. Foundation for Meteoritics and Planetary Science team member. A space content creator who is passionate about rockets.

IG: *@thegalacticraven*
TT: *@thegalacticraven*
X: *@RavenNephtis*
FB: *Oumaima Ravens Art and Astrobiology*
YT: *@astrobiologistraven*
LI: *Oumaima Mouttaky*

OUMAIMA MOUTTAKY, aka Astro Raven, a Space enthusiast making waves from Morocco.

She started her journey as a lab tech and later became an integral part of a foundation focused on Meteoritics and Planetary Science. Graduating in Fundamental Biology with a project of an extraordinarily resistant bacterium called Deinococcus Radiodurans, that withstands harsh environmental conditions present in outer space.

Currently, she's transitioning from research to pursue a Master's in Bioengineering, with a focus on Medical, Agri-Food, and Environmental applications.

Oumaima aspires to embark on new experiences within the space industry, while also encouraging young minds to explore STEM fields. She eagerly participates in STEM events, where her enthusiasm shines as she shares captivating stories about meteorites and their importance in geological studies, in uncovering the early conditions and processes of the solar system. Additionally, she enthusiastically engages audiences with discussions on rockets and space industries, sparking curiosity and fostering a deeper appreciation for these cutting-edge fields of exploration as a scientific mediator.

Oumaima also showcases her creativity in a fun way as a space content creator on social media @thegalacticraven. including Q&A sessions with individuals in the STEM field through Youtube: Astrobiologist Raven.

STEM PROs Questions

Why do you do what you do? / Do you have a defining/ah-hah/ eureka moment where you knew what you wanted to do?

Growing up, I was captivated by the mysteries of space, with a deep fascination for NASA documentaries and the breathtaking view of rocket launches. However, it wasn't just my exposure to these documentaries that fueled my passion—but also by the support of my Dad, who nurtured my passion by bringing home books about space and cosmos and watching with me more NASA documentaries whenever we both had the chance. without forgetting one of my most cherished memories, lying on our old house's rooftop, mesmerized by the starry sky and the commanding presence of the Moon overhead. These moments of wonder remained etched in my mind and heart.

So, as I progressed through my studies in biology and now engineering, I discovered inspiring stories of astronauts like Kate Rubins, who shattered barriers and exemplified the limitless potential of human achievement. It became clear to me that my background in science could seamlessly intersect with my passion for space exploration. Kate Rubins' journey, from being a biologist to an astronaut conducting experiments aboard the International Space Station, showcased the incredible opportunities for scientists to contribute to space exploration. Her story resonated with me deeply, reaffirming my belief that I could do something similar as well, while leveraging my scientific background. same for female astronauts of previous generations.

I can also recall a defining 'ah-hah' moment when I realized that my childhood fascination with space was more than just a fleeting interest—it was a calling. It was the moment I recognized that my passion for science and engineering were true. And that I wanted to be part of something bigger than myself, to push the boundaries of human knowledge and exploration. Also is it not just about fulfilling a childhood fantasy, whether in or outside the space industry in another domain, but it's about embracing the limitless possibilities of the universe and inspiring future generations to reach for the stars.

What is something you wish you knew while in school?

One thing I wish I knew while in school is the importance and value of being a polyglot. I've always admired individuals who are fluent in multiple languages and can effortlessly communicate across different cultures and communities. As our world becomes increasingly interconnected, proficiency in multiple languages has become a highly desirable skill, not only for personal enrichment but also for professional opportunities. And I'm working on that, as I already speak Arabic,French and English. so aspiring for more.

What is something you wish you did differently when you first started working?

When I first started working, I wish I had addressed my imposter syndrome head-on. I often doubted my abilities and felt like I didn't belong in my role, despite receiving positive feedback. If I could go back, I would have sought support from mentors or colleagues to overcome these feelings earlier and realized that I was more capable than I gave myself credit for, no matter what the results.

If someone wanted to do what you do, what's the best piece of advice you'd share with them?

If someone wanted to pursue a similar path to mine, I would share the importance of perseverance in the face of obstacles. My own journey has been marked by numerous challenges and transitions. I started with studies in economics before realizing my passion lay in biology. It was after being inspired by the biologist Kate Rubins that I began dreaming of doing the same experiments aboard the International Space Station (ISS). Sadly I found myself grappling with stereotypes and societal expectations that seemed to limit my aspirations, especially considering the lack of established space industries in my area.

However, I later came to realize that every experience is valuable and that it's never too late to pursue your passions. And that everything happens for a good reason, and this is leading me to appreciate the diverse journey I had undertaken.

If you could definitively answer one unanswered science question, what would it be and why? / What unanswered scientific question keeps you up at night?

If I could answer one big science question, it would be how big the universe really is. This is fascinating because it could tell us if there's more out there than we ever imagined. It's like finding out just how vast our cosmic neighborhood truly is.

*Still the same wonder keeps me up at night: how big the universe really is. It's a mystery that sparks endless curiosity and imagination. Just pondering the vastness of space and the potential for what lies beyond our current understanding leaves me wide awake with wonder. Despite our advancements in exploration and technology, there's so much we have yet to uncover.

What scientific discovery or event in history would you like to go back in time and witness?

One of my favorite questions! If I could go back in time to witness any scientific discovery or event, I would choose to witness the first landing of astronauts on the moon. Just imagine the exhilaration and awe of being one of those astronauts, taking that historic step onto the lunar surface and proving to the world that it's not just a dream—it's reality. The sheer joy and sense of accomplishment in that moment would be unparalleled.

Additionally, I would love to witness the launch of the space shuttle. The sight of that magnificent spacecraft soaring into the heavens, carrying astronauts on missions of exploration and discovery, is a truly breathtaking moment. I find myself feeling a twinge of jealousy towards those fortunate individuals who had the opportunity to witness such a remarkable event firsthand. They were truly lucky to witness history in the making

What's a misconception about your line of work?

In the science field, such as biology, there's often a misconception that scientists or technicians only work in laboratories conducting experiments. While laboratory research is a significant aspect, scientists also engage in fieldwork, data analysis, teaching, and science communication. They may conduct experiments in diverse settings, collaborate with interdisciplinary teams.They can also take on leadership positions within their organizations. They may serve as entrepreneurs, managers, department heads, or project leaders, overseeing research initiatives, guiding teams of scientists and technicians, and setting strategic directions for the organization. Their leadership skills and scientific expertise are instrumental in driving innovation.

Another common misconception about being a Bioengineer for example in the space sector is that it's limited to traditional engineering roles and doesn't involve biology or life sciences. However, bioengineers play a crucial role in space exploration by leveraging their expertise in both engineering and biology to develop innovative solutions for astronauts' health and well-being in space. And not only coding or building machines.

What's the best part of doing what you do?

The best part of what I do is the blissful opportunity to maintain a balance between full-time work and full-time studies, even during nights and weekends. It's undeniably challenging and demanding, but it fills me with immense pride to pursue something I've always dreamed of. This unique experience not only tests my perseverance and time management skills but also fuels my passion for continuous learning and growth. Despite the difficulties, I find fulfillment in knowing that I'm actively working towards my goals and taking steps towards a future that excites me.

What is a necessary evil in your industry?

One necessary evil in any industry is the presence of individuals who feel superior in their positions. It's something I dislike. As this can lead to power struggles, stifling collaboration and innovation. It creates a negative work environment where some voices are valued more than others. However, by promoting humility and inclusivity, we can work towards creating a more positive and productive workplace culture.

As a future leader, I dislike seeing people act superior. I believe in treating everyone equally. Despite this behavior in the industry, I'm committed to promoting fairness and inclusivity. It's important to create an environment where everyone feels valued and respected, regardless of their position.

Have you ever changed your line of work? If so, why and what was the change?

Yes, I have changed my line of work multiple times. I made these changes because I refused to accept being treated as inferior or succumbing to stereotypes. Battling imposter syndrome, I realized that staying in environments that didn't value my contributions was detrimental to my well-being and growth. Therefore, I sought new opportunities where I could thrive and be recognized for my abilities and potential. Each change allowed me to pursue paths aligned with my values and aspirations, leading to greater fulfillment and success in my career journey

What purchase under $100 has improved your life: career or personal?

While it's hard to pinpoint a specific purchase under $100 that has transformed my life, I vividly recall investing in books and magazines to fuel my curiosity and thirst for knowledge. Delving into subjects like self-growth, science, space, mythologies, and history has been a constant source of inspiration and learning for me. These resources have not only expanded my understanding of the world but also fueled my personal and professional growth, allowing me to cultivate new perspectives, insights, and passions along the way.

How has social media benefited or hindered your career?

Social media has been a mixed bag for my career. On one hand, it has allowed me to connect with professionals in my field and learn from their experiences and build close relationships within the space community, making it feel like a family, particularly those who have worked at NASA and other space-related organizations. These connections have been invaluable for gaining insights into career trajectories and landing opportunities in the industry. Additionally, engaging with space-themed communities has led to enjoyable collaborations and projects that align with my interests.

STEM PRO
ATHENA BRENSBERGER

ATHENA BRENSBERGER: Astronomy BS in progress, Astrophysics/ Space Content Creator, Host & Science Communicator, Model

IG: *@astroathens*
TT: *@astroathens*
X: *@astroathens*
YT: *@astroathens*

ATHENA BRENSBERGER is a passionate science communicator and astronomer aiming to make science more accessible and exciting for everyone. Athena has a space segment 'Astro Athena' on *CBS Mission Unstoppable Season 5*, she is the host of the TV show *Suppressed Science*, is a former researcher at the Hayden Planetarium/AMNH, a current student at Arizona State University, and founder of Astroathens where she creates educational science content. She aims to expand the reach of STEM to kids worldwide, specifically to increase the representation of girls in various STEM fields!

STEM PROs Questions

Why do you do what you do? / Do you have a defining/ah-hah/ eureka moment where you knew what you wanted to do?

I found my ah-ha moment in a 13-year-old girl aspiring to become an astronaut. She is years ahead of where I was when I was her age. When she met me said she'd looked up to me for years—my entire energy shifted in this moment. Here I was meeting someone who had known about me since she was a very young girl, and sharing my story and journey in STEM, inspired her to pursue a similar path. I knew why I started creating content around space at the beginning, but when I met her, that reason became alive in my reality. A small amount of what I put out might be just enough for one person to start an everlasting ripple in their life and the world. It's not about me, it never has been, and I just love that.

What is something you wish you knew while in school?

There's no need to rush *anything*.

If someone wanted to do what you do, what's the best piece of advice you'd share with them?

Just start. You might stumble along the way at the beginning but eventually, you will stand up straight. No one starts anywhere as an expert, you just have to get started and the confidence will come as a result of learning and struggling along the way.

If you could definitively answer one unanswered science question, what would it be and why? / What unanswered scientific question keeps you up at night?

Multiverse/beginning of the universe (pre-big bang) and life beyond Earth. These two things have kept me up at night outside under the stars in my backyard thinking for hours. How can I feel so connected to this void, yet not have concrete answers of knowing all that's out there?

What scientific discovery or event in history would you like to go back in time and witness?

It would have to be between the discovery of the Moons around Jupiter and the discovery that the Andromeda Nebula was the Andromeda

Galaxy. These two INSANE shifts in reality rapidly launched us forward in consciousness, where we began to gain true awareness of where we are in space and realize that it's vastly different than what we previously thought.

What's a misconception about your line of work?

That I know all the answers about space lol no one does.

What's the best part of doing what you do?

Connecting to people.

Have you ever changed your line of work? If so, why and what was the change?

I was a fashion model before I started Astroathens and began communicating science, this was while I was in my first years of Undergrad. I have since changed my career to being a part-time content creator, TV Host for science shows, then a part-time barre fitness instructor, and now a full-time content creator as I complete my astronomy degree online where I will likely change my career again to be a research astronomer.

What purchase under $100 has improved your life: career or personal?

I had to think about this one but my mind kept coming back to my external hard drive, LaCie $95 on sale. It changed the game for me storage-wise and for video editing since I edit on the hard drive rather than on the computer's storage. I used to have computer crashes all the time from never having enough storage.

How has social media benefited or hindered your career?

Social media made my career happen in the first place. Without it, I would be pursuing a very different career path. It opened the way for me to make it to television as a science host and connect with people through my various channels. I learned skills I didn't have before starting Astroathens on social media, such as communication skills, video editing, how to use professional camera equipment, topic research, and scriptwriting. I truly owe it to social media for being able to create what I want to do in the world, plus social media has become the perfect place to showcase a resume in media form.

STEM PRO

JENNA P. MERCURIO

JENNA P. MERCURIO: BS Biology, wildlife & nature photographer, safari host & animal care giver.

IG: *@jennamercurio*
Website: *jennapricephotography.mypixieset.com*

JENNA P. MERCURIO was born and raised in New York and still currently resides there. She has a BS in Biology with a concentration in ecology and evolution from Stony Brook University. During her college years she studied abroad in Tanzania which completely changed her life.

Currently Jenna is an animal caretaker, wildlife photographer and safari host. Her greatest passion is being able to travel the world and photograph wildlife and the beautiful areas they call home. Through her photography she hopes to inspire others to learn more about the natural world and ways to conserve and protect it. She is honored to also be able to host safaris in Tanzania to get people closer to the wilds of Africa for their own life changing experience.

STEM PROs Questions

Why do you do what you do? / Do you have a defining/ah-hah/ eureka moment where you knew what you wanted to do?

Ever since I was a kid I always had a love for wildlife and the natural world. There was never a doubt that working in a biological field would be my calling. As I grew up, I also developed a passion for photography. I visited a National Geographic photo gallery about 5 years ago which really inspired me to focus on and take my art to the next level. I remember feeling emotional looking at some of the images and knew I wanted to be able to do the same with my photography.

What is something you wish you knew while in school?

In school I wish I was less harsh on myself and looked into more experiences/ volunteering/internships. Getting out and seeing the world and partaking in projects that interest you is more important and valuable than textbooks and tests, in my opinion.

If someone wanted to do what you do, what's the best piece of advice you'd share with them? and What's a misconception about your line of work?

There is a big misconception that being an animal caregiver entails just playing with animals all day. Taking care of animals is one of the toughest jobs. The pay is very low, it's physically, emotionally and mentally taxing, you work weekends and holidays, miss out on big events, and there are a lot of people out there that will not support you or the facilities you work in. There is a very large chance you will also suffer from compassion fatigue. In short, you give a lot of yourself and barely get anything, including a livable wage, in return. These are things anyone looking to be in this field should know, and should know how to maintain a healthy work/life balance.

If you could definitively answer one unanswered science question, what would it be and why? / What unanswered scientific question keeps you up at night?

I would love to know exactly what dinosaurs looked like!

What's the best part of doing what you do?

Best part of being an animal caregiver is teaching guests about wildlife and threats to their wild populations and inspiring them to protect and conserve them in whatever ways they can. The best part of being a wildlife photographer and safari host is being able to share my life changing experiences abroad through my photography or in person on my safaris.

How has social media benefited or hindered your career?

Social media has been a huge benefit to my life overall. Because of it, I was offered the opportunity to be a safari host which is truly a dream come true. It's also been amazing connecting with people from all of the world who share the same passions of wildlife conservation and photography.

STEM PRO

SONIA CAMACHO

SONIA CAMACHO: MEng Computer Science, BS Computer Science Systems, Nike Engineer

IG: *@sonia_macho*
TT: *@sonia_macho*

SONIA CAMACHO obtained a B.S in Computer Science Systems (2021) and a Masters of Engineering in Computer Science in (2022). Both degrees were completed at Oregon State University. In Sept. 2022 Sonia started a full time position as an engineer at Nike.

Throughout her academic career Sonia began documenting her passion and struggles as a Latina woman in STEM. Through social media, Sonia has been able to grow her online presence to inspire and encourage the next generation of underrepresented students.

STEM PROs Questions

Why do you do what you do? / Do you have a defining/ah-hah/ eureka moment where you knew what you wanted to do?

I am a software engineer because I love solving problems and helping others.

What is something you wish you knew while in school?

I wish I knew how to effectively learn with ADHD. In grad school I took a psychology class that was all about the science of learning. It wasn't until then that I learned different learning styles and how to best study as a student. Learning how to learn may sound silly but it makes a world of a difference for comprehension and retainment of information.

What is something you wish you did differently when you first started working?

I wish I asked more questions. I am finally at a point where I feel confident in my skills to ask questions and own what I don't know. At first it can be embarrassing to ask questions but once you get over the perfectionist mindset and shift to a learning mindset you really open the doors to learning. By learning more you will advance further in your career.

If someone wanted to do what you do, what's the best piece of advice you'd share with them?

Learn soft skills like networking and working together in a team. Anyone can learn to code, but fostering a good community and team environment is hard to do. These skills are often overlooked in this field but they are key for success and building connections.

If you could definitively answer one unanswered science question, what would it be and why? / What unanswered scientific question keeps you up at night?

I can't help but wonder if there are other universes? Or is our Universe the only one?

What scientific discovery or event in history would you like to go back in time and witness?

The invention of the phone would have been cool to witness. We have the world at our fingertips in our smartphones but I wonder if the very first phone calls must have felt unreal to hear someone's voice on the other side from a different location. This invention is interesting because it has evolved so much!

What's a misconception about your line of work?

If you weren't born a math genius you will fail. I struggled a lot in school and had multiple tutors to help me grasp basic concepts. With enough drive and motivation you can succeed in any subject in school. I think this misconception drives a lot of people away from entering computer science which is sad!

What's the best part of doing what you do?

Helping others! Everyday I work on bug fixes that help our consumers especially those with disabilities and knowing that I am improving someone's shopping experience is a great feeling. I didn't learn about accessibility much in school and looking back I wish it was talked about more, because accessibility is so important in any business!

What is a necessary evil in your industry?

I guess I would say pressure. Not everyone works well under pressure but I thrive in being booked and busy. Having pressure or deadlines help motivate me to get my work done and get it done right.

Have you ever changed your line of work? If so, why and what was the change?

I have not! But recently I started taking criminal justice classes but just for fun!

What purchase under $100 has improved your life: career or personal?

Planner—time management is KEY!

How has social media benefited or hindered your career?

Social media has helped my network tremendously! I love posting my daily life and encouraging others to pursue STEM. Through this I have made so many connections at work and across the country. I have been a part of some really cool projects and had amazing opportunities because of my social media. The best part of social media is talking to students and giving them advice as they prepare for their career! I love sharing my knowledge with the next generation of engineers!

STEM PRO

MICHELLE TOVAR-MORA

MICHELLE TOVAR-MORA, MSME, EIT: Mechanical Engineer in the Energy Sector, Women in STEM and Working Moms Advocate

IG: *@galvanizestem*
TT: *@galvanizestem*
LI: *Michelle Tovar-Mora*

MICHELLE TOVAR-MORA, a trailblazer in the engineering field, embodies the spirit of perseverance and advocacy. Born to Mexican parents, Michelle grew up with a deep appreciation for hard work and determination instilled by her family's values. As a first-generation college student, she defied the odds to pursue her passion for engineering. Her journey began with earning her Bachelor of Science in Mechanical Engineering, a milestone that marked the beginning of her dedication to the field. Driven by her thirst for knowledge, she furthered her education by obtaining a Master of Science in Mechanical Engineering. Although initially choosing mechanical engineering to pursue a career in aerospace, Michelle embarked on to her professional career in the power sector, where she has dedicated almost a decade to making significant contributions. Her expertise and commitment have propelled her to excel in her role, earning her recognition as a valuable asset in her field.

Beyond her academic and professional accomplishments, Michelle is a devoted mother to three beautiful children, balancing the demands of motherhood with her career aspirations. As a first-generation engineer, advocate, and mother she exemplifies resilience, determination, and compassion.Till this day Michelle remains deeply connected to her cultural heritage and roots and takes pride in her Mexican heritage. By embracing her identity and cultural background, Michelle serves as a role model for aspiring engineers from similar backgrounds, showing them that success in the field is attainable regardless of one's ethnicity or cultural upbringing. Her ability to juggle multiple responsibilities with grace and determination serves as an inspiration to many.

Recognizing the importance of representation, she actively engages in mentorship programs, outreach initiatives, and speaking engagements to inspire and empower individuals from diverse backgrounds to pursue careers in engineering and related fields.As a woman of color and a first-generation engineer, Michelle understands the significance of providing support and guidance to those individuals from underrepresented backgrounds who may face systemic barriers or lack access to resources. She actively seeks opportunities to mentor aspiring engineers, offering advice, encouragement, and networking connections to help them navigate their academic and professional journeys.

Michelle has served on multiple committees, boards, and advisory panels focused on advancing the representation of women, mothers, minorities, and first-generation college students in engineering and technical fields. Through her leadership, mentorship, and advocacy efforts, she continues to pave the way for future generations of engineers, ensuring that the field remains accessible and welcoming to individuals from all walks of life

STEM PROs Questions

Why do you do what you do? / Do you have a defining/ah-hah/ eureka moment where you knew what you wanted to do?

As a Latina engineer working in the power sector, my motivation stems from a combination of personal passion, professional interest, and a desire to make a positive impact on the world around me. Growing up, I didn't always see anyone who looked like me in STEM fields, and I understand the importance of representation in inspiring others to pursue their passions and reach their full potential. As a kid I was always fascinated by how things were built and how they functioned but, I did not realize until I went to college that it was an interest in engineering all along. As I pursued my college education and joined the Society of Women Engineers, I realized I wanted to become an engineer. I became increasingly drawn to the challenges and opportunities that engineering could bring to specially growing up in a low income household. Although I initially studied engineering with an interest in Aerospace One defining moment for me to enter the Energy Sector was getting hired in that field upon completing my Bachelors of Science in Mechanical Engineering. As I delved into the complexities of energy technologies and their potential to mitigate, promote sustainability and optimize power in the most economic and reliable, I experienced a sense of purpose and fulfillment that solidified my commitment to pursuing a career in the power sector. Moreover, as a Latina engineer, I am motivated by the opportunity to contribute to diversity and representation within the engineering field. I believe that diversity fosters innovation and creativity, and I am passionate about breaking down barriers and empowering individuals from underrepresented backgrounds to pursue careers in STEM fields. I am passionate about finding innovative solutions to address the current challenges, improve energy access, and promote social and environmental justice. Through my work as an engineer, I strive to create positive change that benefits not only current generations but also those to come. Overall, being a Latina engineer in the power sector is about more than just a career—it's about making a difference, breaking down barriers, and inspiring others to dream big and pursue their goals, no matter their background or circumstances.

What is something you wish you knew while in school?

As a first-gen Latina engineer, I wish I knew more about the various support networks and resources available for underrepresented groups in STEM fields while I attended college. Building connections with mentors and peers who share similar experiences can be incredibly valuable for navigating challenges and finding opportunities for growth and success. Additionally, I wish I had known more about the importance of self-advocacy and confidence in pursuing my career goals, as these skills are essential in a competitive field like engineering.

What is something you wish you did differently when you first started working?

Looking back, I wish I had been more proactive in seeking out mentors and allies in the workplace who could offer guidance and support as a first-generation Latina engineer. Building a strong network of colleagues and mentors early on can provide invaluable advice and help navigate the challenges that come with being underrepresented in the field. Additionally, I wish I had been more assertive in advocating for myself and my ideas, as confidence and self-advocacy are crucial for success in any career, especially in engineering.

If someone wanted to do what you do, what's the best piece of advice you'd share with them?

The best piece of advice I would share with someone who wants to pursue a career similar to mine as an engineer is to believe in yourself and your abilities. Recognize that your unique perspective and background bring valuable insights to the field. Stay curious, continue learning, and never underestimate the power of perseverance in achieving your goals. In addition embrace challenges as opportunities for growth: Engineering can be complex and demanding, but every challenge you face is a chance to learn and develop new skills. Don't shy away from difficult tasks; instead, approach them with determination and a willingness to learn. Connect with colleagues, mentors, and industry professionals who can offer guidance, support, and opportunities for career advancement. Also remember that networking is key to success in any field, and it's especially important for first-generation professionals who may not have access to established networks through family connections.

If you could definitively answer one unanswered science question, what would it be and why? / What unanswered scientific question keeps you up at night?

As a woman with Polycystic Ovary Syndrome (PCOS, one unanswered scientific question that is particularly significant to me is finding the exact cause or causes of the condition. PCOS is a complex hormonal disorder that affects reproductive-aged women and is characterized by a combination of symptoms such as irregular periods, excess androgen levels, and polycystic ovaries. While researchers have made progress in understanding the mechanisms underlying PCOS, the exact etiology remains unclear. There are likely multiple factors at play, including genetics, insulin resistance, hormonal imbalances, and environmental influences, but the precise interplay and causal relationships are not fully understood. I believe that just like any problem, finding a definitive answer regarding the root causes of PCOS could lead to more targeted and effective treatments, as well as strategies for prevention and early intervention. It could also help alleviate the physical and emotional burden that PCOS imposes on millions of women worldwide. As a woman with this syndrome and engineer, this unanswered question is particularly compelling because of its potential to improve the lives of so many individuals like myself and families affected by PCOS. It's a question that drives ongoing research efforts and fuels my curiosity about the complexities of women's health and reproductive biology.

What scientific discovery or event in history would you like to go back in time and witness?

As an engineer in the Energy sector the one event I would love to go back in time and be part of is definitely the Enron scandal which not only had a significant event in the energy industry it has had lasting impacts on the sector. Enron Corporation, was once one of the largest energy companies in the world; it collapsed in 2001 due to widespread accounting fraud and corporate misconduct. At the heart of the scandal was Enron's use of complex financial schemes and accounting practices to manipulate its financial statements, conceal debt, and inflate profits. This deceptive behavior misled investors, regulators, and employees, resulting in billions of dollars in losses and the eventual bankruptcy of the company. The Enron scandal led to increased scrutiny and regulation of the energy industry, as well as reforms aimed at improving corporate governance, transparency, and accountability. It also highlighted the risks of excessive deregulation and the need for effective oversight to prevent fraud and abuse in the marketplace. Overall, the Enron scandal serves as a cautionary tale about the dangers of unethical behavior and the importance of integrity, transparency, and accountability in the energy industry and beyond. It continues to shape discussions around corporate governance, risk management, and regulatory oversight in the modern business world. Although one of the most controversial events in the energy sector, I would have loved to experience this first hand instead of just reading about it and hearing all of the stories because till this day my job functions have been impacted by this one event in history.

What's a misconception about your line of work?

One common misconception about the energy sector is that it is not innovative and it is resistant to change. However, the reality is that the energy industry is undergoing rapid transformation, driven by technology advancements, policy changes, and evolving customer preferences. Although traditional fossil fuel-based energy sources, such as oil, gas, and coal are still needed for reliability on the grid, other significant components of the energy sector are a growing emphasis on renewable energy sources like solar, wind, hydroelectric, and geothermal power. Many engineers in the energy sector are actively involved in developing and implementing sustainable energy solutions, including designing more efficient solar panels, optimizing wind turbine technology, improving energy storage systems, and enhancing grid infrastructure to accommodate renewable energy integration, grid technologies, and electric vehicles which are reshaping the way we produce, distribute, and consume energy. Additionally, the energy sector encompasses a wide range of disciplines beyond engineering, including policy development, finance, project management, and environmental science. Engineers in the energy sector often collaborate with professionals from diverse backgrounds to address complex challenges related to energy production, distribution, and sustainability. Being an engineer in the Energy Sector brings ample opportunities to contribute to the transition towards a more sustainable and environmentally friendly energy future.

What's the best part of doing what you do?

One of the best parts about being a bilingual Latina engineer in the energy sector is the ability to bridge cultural and language barriers in a diverse and global industry. Being bilingual allows you to communicate effectively with colleagues, clients, and stakeholders from different backgrounds, fostering collaboration and understanding across linguistic divides. Additionally, being Latina brings a unique perspective to the table, enriching discussions and problem-solving processes with insights informed by your cultural heritage and experiences. Your bilingualism and cultural background can also open doors to opportunities for international work or projects that involve engagement with Spanish-speaking communities. Overall, being a bilingual Latina engineer in the energy sector offers the chance to leverage your language skills and cultural identity to make meaningful contributions to projects, teams, and initiatives, while also promoting diversity, inclusivity, and cross-cultural understanding in the workplace.

What is a necessary evil in your industry?

One necessary evil in the power industry is the reliance on certain forms of energy generation that are perceived to have negative environmental impacts. For example, while fossil fuels like coal and natural gas are currently essential for meeting energy demand, their combustion releases greenhouse gasses and pollutants that contribute to climate change and air pollution. Despite efforts to transition to cleaner and more sustainable energy sources, such as renewable energy and nuclear power, these alternatives may not yet be capable of meeting all energy needs reliably and affordably on a large scale. As a result, the power industry must balance the imperative to reduce environmental harm with the need to ensure a stable and affordable energy supply. This tension between environmental sustainability and energy economics and reliability is a necessary evil that the power industry grapples with as it strives to meet growing energy demands while mitigating the negative impacts of energy production, consumption on the planet, and affordability impacts to consumers.

What purchase under $100 has improved your life: career or personal?

A gym membership to keep me both physically and mentally sane.

How has social media benefited or hindered your career?

As a first-generation Latina engineer, I truly feel that social media can offer numerous benefits to support your career and personal development. For me personally, the way I utilize social media platforms has allowed me to not only benefit in my engineering career but also in motherhood. Building a strong online network has opened doors to career opportunities, collaborations, and professional growth. Social media allows me to stay informed about industry news, trends, and advancements in engineering. By following relevant accounts, groups, and pages, I have been able to access valuable resources, articles, webinars, and discussions that have enhanced my knowledge and skills. It has also enabled me to showcase my engineering expertise, achievements, projects, and motherhood journey to a broader audience. By sharing my work, insights, and experiences, I have been able to establish myself as a thought leader and build a strong personal brand within the engineering and working mothers community. Being a first-generation Latina engineer while navigating being a mother to three small children can sometimes feel isolating, but social media provides a platform to connect with others who share similar backgrounds and experiences. A lot of the online communities, groups, and forums have offered support, encouragement, and mentorship from peers and role models navigating similar paths. In addition being a first-generation Latina engineer and mother, social media has allowed me to advocate for diversity, inclusion, and representation in STEM fields and working mothers groups. By sharing my story, amplifying diverse voices, and raising awareness about important issues, I have been able to contribute to a more inclusive and equitable engineering, stem, and working communities which I would not have otherwise connected with. Overall, social media can be a valuable tool for empowering first-generations, women in STEM, and mothers, providing opportunities for networking, learning, visibility, and advocacy that can support your career success, motherhood success, and personal growth.

However, I do believe that due to the political, ethical, and professional concerns one needs to be mindful when posting on personal social media platforms to not hinder their own career.

STEM PRO

ERIN WINICK ANTHONY

ERIN WINICK ANTHONY: Science Communicator| Founder @STEAMPowerMedia | Formerly NASA, MIT | BS Mechanical Engineering

IG: *@erinwinick*
Threads: *@erinwinick*
TT: *@erinwinick*
X: *@erinwinick*
FB: *Erin Winick*
YT: *@erinwinickanthony*
Websites: *erinwinick.com, steampowermedia.com*

ERIN WINICK ANTHONY is passionate about showing the art and creativity in science. She founded her science communication company, STEAM Power Media, after nine years working as a professional science communicator for places like NASA and MIT. Her company helps everyone from scientists to engineering companies to museums explain complex technical topics through writing and social media. She even sailed on a two month ocean expedition as an outreach officer for the JOIDES Resolution, sharing the geological research of the mission.

At NASA Erin worked as a science communications specialist for the International Space Station. Based out of Johnson Space Center, she helped tell the stories of research conducted aboard the International Space Station through writing and social media. She helped create the next edition of the space station Benefits for Humanity Book, served as a producer for the NASA Explorers: Microgravity video series, managed the @ISS_Research Twitter account, and more.

Erin has a B.S. in mechanical engineering from the University of Florida. She has also written and worked for MIT Technology Review and The Economist. In her free time she plays competitive pinball, loves 3D printing and creating science inspired fashion. Yes, her closet does look like Ms. Frizzle's.

STEM PROs Questions

Why do you do what you do? / Do you have a defining/ah-hah/ eureka moment where you knew what you wanted to do?

At my core, I have always loved making things. Whether that was using a lathe to make a part I designed on the computer or building a Rube Goldberg machine in my parent's living room, making has always been a part of who I am. That is what first led me to engineering.

However, I found my niche when I realized that so many engineers around me didn't enjoy explaining these complex engineering topics. There is a need for better STEM communicators and as someone with a passion for writing, I found my niche.

What is something you wish you knew while in school?

There are so many more career paths out there than you realize. You can use your unique set of skills to even form your own career path! I used to think the only science communicators out there either wrote technical manuals or were hosts of TV shows like Bill Nye. I have enjoyed carving my own path in this space, but wish I had been able to work in this space sooner.

What is something you wish you did differently when you first started working?

I wish I had documented the early stages of my career more and shared more honestly what I was working towards. It's hard to share when you are vulnerable and unsure, but being more open about your goals allows more people to help guide you in the right direction.

If someone wanted to do what you do, what's the best piece of advice you'd share with them?

Start sharing what you are learning right now. Science communication is a skill you build up over time, but you don't have to wait until you get a PhD to start communicating science. Share fun facts you learn at school with your friends and family. Offer to write articles for smaller blogs or publications to get practice spreading the word about science.

If you could definitively answer one unanswered science question, what would it be and why? / What unanswered scientific question keeps you up at night?

I'm going to cheat and give two answers here.

1. Are we alone in the universe and if not, where is that life? The eternal science question for those of us who love space, I'd love to know where we should be looking in our universe for alien life!
2. At this point, what is the best-case scenario for curtailing climate change? Humanity continues to emit more greenhouse gasses and the problem only seems to be getting more pressing. I want to know just how bad this is going to get, and truthfully what we can do to stop it at this point.

What scientific discovery or event in history would you like to go back in time and witness?

I would love to go back in time and witness the first ever Moon landing. So many people I worked with at NASA cited that moment as the reason they first got into studying space. It must have been incredible to have the whole world tuned into this moment focused on space exploration.

What's a misconception about your line of work?

It is a bit hard to explain how what I do turns into a full time job. I work on so many individual different projects for many different clients. One day I am doing consulting with a water testing company, the next I am filming videos for an environmental startup.

But the most common misconception is that all science communicators are essentially Bill Nye, working on recording TV shows for kids.

What's the best part of doing what you do?

I love constantly learning every day, and working directly with the scientists who are making groundbreaking discoveries. I got the opportunity to sail on an ocean expedition that was using rocks pulled from beneath the ocean floor to study Earth's past climate. It was incredible to be out in the field where the research was happening and share that story online.

What is a necessary evil in your industry?

Charging money to organizations that I wish I could support for free. It sometimes feels wrong to pitch high amounts for projects with nonprofits or museums, but it's always a balance between supporting myself and working to share science that I care about.

Have you ever changed your line of work? If so, why and what was the change?

Yes! I studied mechanical engineering, and debated getting an engineering job or taking the risk and trying to venture into science communication. After having four engineering internships, I just felt that my passion was combining this technical background with my love of communications.

I did also shift from working in a full-time role to running my own company and working on freelance science communication opportunities. I'm someone who loves working on 15 different things at once, and having the freedom to chase after unique opportunities. This career change enabled that.

What purchase under $100 has improved your life: career or personal?

OneOdio corded over-the-ear headphones. I have extremely sensitive ears and always struggle with them hurting after wearing almost all headphones for long periods. These were a lifesaver and are my go to for editing on the computer and even listening to music during pinball tournaments.

How has social media benefited or hindered your career?

It has benefited my career so much. I have gained so many connections, friends, and job opportunities because of the content I have created online. Being a visible face in the STEM fields helps show people where your priorities are at, serves as a portfolio for your work, and allows you to be a role model for the next generation!

STEM PRO
NAOMI THOMAS

NAOMI THOMAS: Tech Entrepreneur

IG: *@techbaenae*
X: *@techbaenae_*

NAOMI THOMAS is an entrepreneur and advocate for diversity in tech. She is currently the Head of Digital at Stemuli, a gaming company at the intersection of AI, education, and workforce development. Naomi previously founded infinity.careers, an award-winning tech career exploration and talent training marketplace, acquired by Stemuli in April 2023. Naomi started her journey in tech at 6 years old when she built her first computer. She went on to participate in Computer Science Institutes at UC Berkeley and UVA in high school and worked with Google as a student recruiting liaison for a year in college. After graduating, she founded and scaled a digital marketing agency to a 30+ creative team and global clientele. Naomi is passionate about using her voice as a guide to encourage other women to fearlessly pursue multi-faceted careers.

STEM PROs Questions

Why do you do what you do? / Do you have a defining/ah-hah/ eureka moment where you knew what you wanted to do?

My journey in tech started at six years old when I built my first computer. As I navigated through Computer Science Institutes at UC Berkeley and UVA during high school, and later as a student recruiting liaison with Google in college, each step further solidified my passion for the field. After college, I channeled this passion into founding and scaling a digital marketing agency, where I led a team of over 30 creative professionals, catering to a global clientele. My background in computer science, combined with business management, opened up a world of opportunities. I've had the privilege of spearheading various tech-driven initiatives, from hosting STEM workshops for students to managing mobile and web app development projects for startups. At the core of my journey is a deep-seated passion for empowering young women in STEM. My mom is my inspiration and also a woman in STEM as the Director of Anesthesiology. Without her support and guidance, I wouldn't have had such an early start in the field. Using my experiences as a compass, I aim to influence the next generation of female tech pioneers. It's not just about advancing technology; it's about shaping a future where women and minorities all have an equal stake and voice. That's the driving force behind everything I do.

What is something you wish you knew while in school?

I wish I had known that it's perfectly okay to ask for help. School isn't just about acing tests; it's a journey of learning, stumbling, and getting back up. I would have told my younger self to ease up on the pressure and embrace each stumble as a step forward. Learning to give myself more grace would have been a game changer.

What is something you wish you did differently when you first started working?

I'd tell my younger self: "Naomi, slow down and breathe." I was always in a rush to achieve. Sometimes, the best insights come when you're not pushing hard, but rather when you're taking a moment to just be.

If someone wanted to do what you do, what's the best piece of advice you'd share with them?

Be unapologetically you. Blend your passions with your skills. Tech isn't one-size-fits-all. It's diverse, it's dynamic, and it's ready for your unique touch.

If you could definitively answer one unanswered science question, what would it be and why? / What unanswered scientific question keeps you up at night?

I wish I could fast forward 20 years from now to see what the future of education looks like. I imagine a world where education is fully gamified, personalized, and tightly interwoven with career pathways. How amazing would it be to create learning experiences that truly prepare the workforce with the skills they need, while also helping learners save time and money? That's a future I want to help shape.

What scientific discovery or event in history would you like to go back in time and witness?

The moment the internet went live. To be there as the first digital signal sparked to life, initiating a network that would eventually connect the entire world, is an extraordinary thought. It's not just about the technology, but about the dawn of a new era in human communication and knowledge sharing.

What's a misconception about your line of work?

That building technology is all coding and algorithms. It's so much more. It's about storytelling, creativity, understanding human needs, and designing products for all end users, not just some.

What's the best part of doing what you do?

Seeing the spark in someone's eyes when they realize tech isn't just a field—it's a canvas for their dreams.

What is a necessary evil in your industry?

Deadlines. The perfectionist in me has had to learn to trust the process and understand that building technology is an iterative process—things will never be perfect, but it's difficult to make improvements without testing things out.

Have you ever changed your line of work? If so, why and what was the change?

From digital marketing to building AI powered technology at the intersection of education and gaming—it was a leap, but one fueled by a desire to explore how tech shapes our learning and working. My passions have also helped me become more multifaceted in my endeavors as I now also model and manage my brother's music career. It's all about evolving and following where your curiosity leads.

What purchase under $100 has improved your life: career or personal?

My iPhone's magnetic wallet has been a total game changer. I personally don't like taking a purse everywhere, so it's made navigating the world a little less heavy.

How has social media benefited or hindered your career?

It's been a double-edged sword. On one hand, it's an incredible platform for sharing ideas and connecting with my STEM community and loved ones. On the other, it's a whirlwind of keeping up and tuning out the noise. Balance is key.

"Everything is theoretically impossible, until it is done."

–Robert A. Heinlein, aeronautical engineer and science fiction author

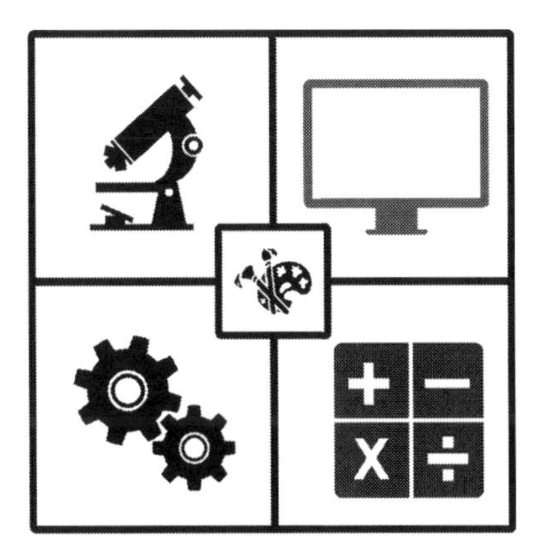

STEM PRO
LANCHEN MIHALIC

LANCHEN MIHALIC: Cosmic Space Artist and Science Communicator

IG: *@lanchendesign*
TT: *@lanchen*
X: *@lanchen*
FB: *Lanchen Designs*
YT: *@justlanchen*
Website: *LanchenDesigns.com*

LANCHEN MIHALIC is a cosmic space artist and science communicator who is passionate about exploring the depths of space and growing the space community.

Ever since she was a kid, she dreamed of being an astronaut and as she got older she found a way to explore space from earth, through her creative expression.

Lanchen fell in love with painting the night sky but more specifically abstract nebula because she felt as if she was painting the deepest parts of space that might exist. Through this passion, Lanchen wants to get more people interested in what connects us all. It's not just about the art, but about growing the space community even further.

STEM PROs Questions

Why do you do what you do? / Do you have a defining/ah-hah/ eureka moment where you knew what you wanted to do?

I just love to create. Bringing my ideas to life in a tangible way and creating art that's impactful to others brings me immense joy. Even if I never shared my work, I would still be creating space art because it fulfills me.

Finding space art specifically, came to me when I was painting a pair of DIY space shoes. At the time I was searching for a niche to create many art forms around, painting, decor, jewelry, furniture, ect. but it was important it was a niche I was passionate about that would be meaningful as well. It wasn't until working on those shoes that I realized space was the perfect niche and fit my criteria. From then on, it became the focus of all my creations and it's been so exciting to grow a space community around space art.

What is something you wish you knew while in school?

That art classes may be an extra curricular, but they can also be a viable career path.

What is something you wish you did differently when you first started working?

I only wish I would've found my niche sooner.

If someone wanted to do what you do, what's the best piece of advice you'd share with them?

I would tell them to do it for the love of creating, not for what creating can get you. If you do it for the potential rewards only, it will be impossible to get through the hard times. I would advise them to never give up. It's cliché I know, but I'm convinced that if you never give up, you are guaranteed to succeed.

If you could definitively answer one unanswered science question, what would it be and why? / What unanswered scientific question keeps you up at night?

I would love to know what happens after death. The idea that once we pass away, we never have the opportunity to experience life again is daunting.

What scientific discovery or event in history would you like to go back in time and witness?

Oh this is hard because there are so many amazing events. If I had to choose, I think it would have been life changing to be on the receiving end of the first telephone invented.

What's a misconception about your line of work?

That art is merely a hobby and can't be a viable career path.

What's the best part of doing what you do?

The best part is that it's something I genuinely love and enjoy doing. Even with the struggles that come from chasing your goals, it's a privilege to struggle doing what you love.

What is a necessary evil in your industry?

Failure. Though difficult, it always pushes you to improve and try new things.

Have you ever changed your line of work? If so, why and what was the change?

Everything I've always done has been creative so I feel not much has changed. Even when I modeled, I was still running an Etsy shop creating jewelry, clothing, furniture, and light fixtures. The only major change was deciding to focus all my creations around space.

What purchase under $100 has improved your life: career or personal?

Epoxy resin! At the time it was the most I had invested into an art project and it's what kickstarted my art journey on Instagram.

How has social media benefited or hindered your career?

Instagram really benefited my art by exposing it to a wide variety of people. Another platform that I find to be extremely beneficial is YouTube. This platform has allowed me to create in a multitude of ways from filming, editing, to room makeovers and more. I find Instagram loves niches which limits my creativity but YouTube provides a space for me to create freely.

STEM PRO
JULANA DIZON

JULANA DIZON: TV Host, Travel & Wildlife Content Creator

IG: *@Julana*
TT: *@Julana*
FB: *Julana Dizon Official*
YT: *@Julana*

JULANA DIZON is a former animal trainer who has merged her passion for animals with her entertainment career in Hollywood. As a television host, she has covered everything from red carpets to fitness events to travel destinations. In her show, Wild Adventures, Julana explores animal tourism experiences and wildlife destinations around the world.

STEM PROs Questions

Why do you do what you do? / Do you have a defining/ah-hah/ eureka moment where you knew what you wanted to do?

The reason I'm so passionate about sharing animal experiences with others is because I love animals! I want to help spread the message of how important it is to coexist with nature, teach travelers about ethical animal tourism, and help raise awareness for conservation organizations. My "ah-hah" moment came a long time ago—back in 2017 when I came up with the idea for my show Wild Adventures. I had been working in Hollywood as a tv host/red carpet host for a few years, and wasn't feeling fulfilled or passionate at all about the celebrity news stories I was tasked with sharing. I wanted to do something that was meaningful to me and helpful for the world, so I created Wild Adventures as a way to showcase destinations and experiences where you can get up close and personal with animals.

What is something you wish you knew while in school?

There are so many more ways to work with animals than just being a vet, an animal trainer, or a marine biologist.

What is something you wish you did differently when you first started working?

If I could go back in time, I would spend a year or so traveling the world and volunteering with different wildlife conservation organizations and animal rescue centers. It would be such a great way to get hands-on experience while seeing the world!

If someone wanted to do what you do, what's the best piece of advice you'd share with them?

Just pick up a camera and practice practice practice. It takes a long time to become confident and comfortable being in front of a camera, but anyone can do it!

If you could definitively answer one unanswered science question, what would it be and why? / What unanswered scientific question keeps you up at night?

Are there more blades of grass or leaves on a tree in the entire world?

What scientific discovery or event in history would you like to go back in time and witness?

The Jurassic Age

What's a misconception about your line of work?

It's all glamorous, all the time. Not true! For every segment or interview you see on tv, there is hours and hours of boring, behind—the-scenes research and preparation taking place.

What's the best part of doing what you do?

Experiencing incredible, once in a lifetime interactions with animals. Also getting to meet some really great people who share my ideals and values and care about protecting our wildlife!

What is a necessary evil in your industry?

Editing and my phone storage always being full.

Have you ever changed your line of work? If so, why and what was the change?

Yes, I used to work hands-on with animals as an exotic animal trainer. I loved the job and the animals, but the hours were around the clock sometimes and the pay was minimum wage. I began modeling part-time which soon turned into full time, and then eventually led me to tv hosting!

What purchase under $100 has improved your life: career or personal?

Personal: The book "Men Are From Mars, Women Are From Venus"

How has social media benefited or hindered your career?

Social media has changed the way I am able to find and connect with other people and places who are working to help save our wildlife. It's been an amazing tool for networking, sharing information, and learning about new opportunities!

STEM PRO

CHRISTOPHER HUIE

CHRISTOPHER HUIE: 1st Jamaican in space, 19th Black Astronaut, Speaker, Leader, Mentor, and Aerospace Engineer with experience in aircraft, spacecraft, helicopters, tiltrotors, and flight simulators

IG: *@AstroChuie*
Website: *astrochuie.com*

CHRISTOPHER HUIE (aka "Chuie") is an astronaut, aerospace engineer, and speaker. Christopher has been with Virgin Galactic since 2016, Richard Branson's human spaceflight company, and has contributed to the company in a variety of impactful ways. His contributions to the emerging commercial human spaceflight industry include six years of leadership in the Flight Sciences External Loads Engineering discipline, supporting both the design and operation of the air-launched Spaceflight System. As a Mission Specialist on Unity 25, he played a key role in Virgin Galactic's final test spaceflight, paving the way for the first commercial spaceflight mission, Galactic 01. A Jamaican American, Christopher is among the first 650 humans to venture into space, the 19th Black astronaut, and the 1st Jamaican astronaut.

Before joining Virgin Galactic, Christopher spent nearly six years at Bell Flight, contributing to various rotorcraft programs as both an External Loads Engineer and a Simulation Engineer. He holds a B.S. in Aerospace Engineering from the University of Maryland, College Park, (UMD) where he was a scholar in the QUEST Honors Program and the Igor Sikorsky Scholarship Program. In 2022, Christopher was inducted into the inaugural class of UMD's Clark School of Engineering's Early Career Distinguished Alumni Society, which recognizes alumni under 40 years old for their leadership, innovation, service, and entrepreneurship.

Christopher's dedication to mentoring and inspiring students in Science, Engineering, and Aerospace is evident through his involvement in Virgin Galactic's outreach initiative, Galactic Unite, and his Co-Founder role in the Black Leaders in Aerospace Scholarship & Training (BLAST) Program. A Virgin Unite 2021 "Together We Can: Change" winner, he is actively involved in outreach efforts to improve preparedness and increase opportunities for underrepresented groups in aerospace. In his free time, Christopher enjoys being a dad, learning, plant-based cooking (and eating!), and pursuing interests in music, rock climbing, Seinfeld, Star Trek, and innovative space business ideas.

STEM PROs Questions

Why do you do what you do? / Do you have a defining/ah-hah/ eureka moment where you knew what you wanted to do?

I'm driven by the thrill of tackling challenges that have never been conquered before. From a young age, I've been captivated by flying machines and the mechanics behind them. The process of transforming an idea into reality is what I find most rewarding. The satisfaction of achieving results fuels my motivation to persist in my efforts.

Looking back, it's clear that my diverse interests as a child were all leading me towards engineering. My aspirations evolved from wanting to be a garbage truck driver (specifically, the ones with robotic arms that pick up trash cans), to an inventor, a Lego designer (I wasn't even sure if that was a real job), an astronaut, and finally, a fighter pilot.

As I delved deeper into the world of aerospace and flight, I made three key realizations that shaped my path:

1. Becoming an astronaut seemed out of reach due to my lack of knowledge on how to pursue it.
2. The path to becoming a fighter pilot involved joining the military, which didn't resonate with me.
3. I developed a fascination for helicopters, particularly the Sikorsky RAH-66 Comanche. Its stealthy, edgy design was, quite frankly, the coolest thing I'd ever seen.

These realizations sparked an insatiable curiosity about how things work. It was then that I decided if I couldn't fly, I would dedicate myself to building things that could.

What is something you wish you knew while in school?

I wish I had understood that asking for help doesn't signify a lack of intelligence. In fact, it's quite the opposite—it's a step towards becoming smarter. I used to believe that if I couldn't figure something out on my own, then I was the problem. As I advanced in my career, I was encouraged to find a mentor, a concept that was entirely new to me. I was familiar with the term 'mentor', but I didn't think I needed one. I believed that asking a mentor for help would be an admission of my inability to solve problems independently, and that I wasn't smart enough.

Fast forward to today, I am now a mentor and a strong advocate for mentorship. I believe that learning from others' mistakes is one of the most valuable skills anyone can acquire, but only after you've mastered learning from your own mistakes; hence the saying, "Know Thyself."

Another thing I wish I had known while in school is how I learn best. We now understand that people have different learning styles, yet our education system is structured to teach all students in roughly the same way. If schools taught students about the process of learning itself, students would be better equipped to tailor their learning journey and advocate for themselves.

Some "learning how to learn" topics I've discovered include: how information is broken down and encoded in the brain; the role our senses play in learning and memory; the learning curve and its counterpart, the forgetting curve; and the impact of emotions, sleep, exercise, and diet on learning. The list goes on. If we want to maximize the performance of any complex system, like our brains, we must understand how it functions.

What is something you wish you did differently when you first started working?

Before I began working full-time, I seriously considered pursuing a master's degree in Aerospace Engineering. The deciding factor to start working after my undergraduate studies was my first employer's tuition assistance program, which would allow me to work and earn my master's simultaneously, free of charge. However, after receiving my first paycheck and experiencing life outside of school for the first time, I decided to postpone my master's degree. I wanted to enjoy the fruits of my labor after being in school for over two decades.

Two years into my full-time job, I started my master's degree but never finished. It was incredibly challenging to return to the student mindset, especially while working a demanding job. If I could change one thing, I would have pursued my master's degree immediately to maintain the momentum of academic life.

If someone wanted to do what you do, what's the best piece of advice you'd share with them?

My career path has been unpredictable and off the beaten path, yet filled with pleasant surprises. The key to my success has been balancing technical proficiency with the ability to communicate complex ideas simply. You can be incredibly intelligent, but if you can't convey your ideas to decision-makers, your brilliance may go unnoticed or have little impact. I really struggled with imposter syndrome, but the advice I've listed below helped me overcome it.

Here's my Top 8:

1. Strive for excellence in every task, regardless of your job, role, or title, whether the task is trivial or challenging.
2. Surround yourself with supportive individuals who celebrate and elevate good work.
3. Combat ignorance by asking questions frequently and learning from others' mistakes.

4. Remember that confidence is built on competence. Confidence without competence is risky in this industry.
5. Reason from first principles and use mental models. Mastering these skills will enable you to solve a wide range of problems and identify opportunities that others might miss.
6. Prioritize progress over perfection and practice discernment. Sometimes, making the least bad decision is the best choice. Indecision often leads to worse outcomes than making the wrong decision. Consider the timing of a decision and its opportunity cost.
7. Recognize that effective communication occurs when there is mutual understanding. Meet your audience where they are, then guide them to where you need them to be.
8. Remember that success lies at the intersection of preparedness and opportunity. You have control over your preparedness. But recognizing opportunities is a shared responsibility between you and your environment. In this way, you can create your own luck.

If you could definitively answer one unanswered science question, what would it be and why? / What unanswered scientific question keeps you up at night?

The question that intrigues me the most is, "Are we in a simulation?" This question leads to the intersection of science and philosophy, which I find both unsettling and fascinating. I often ponder whether our universe is truly deterministic or if it's probabilistic in nature. The dual nature of light as both a particle and a wave is mind-boggling. I'm also intrigued by the power of statistics—how is it that we can predict the future with a specific level of confidence, even without understanding the underlying mechanisms of the thing being predicted?

The existence of quantum mechanics makes me question the concept of "free will," or rather, our perception of it. What's the deal with quantum entanglement? Will faster-than-light travel ever be feasible? These are some of my unanswered questions that keep me up at night.

What scientific discovery or event in history would you like to go back in time and witness?

I'd love to witness firsthand how the Mayans came to invent a new number—a number that literally means nothing. I'd like to see how established civilizations reacted to and ultimately adopted the concept of the number zero. If the debate was anything like that of the heliocentric theory vs. the geocentric theory, I'm sure there would be some fascinating stories to hear.

What's a misconception about your line of work?

In my career as a loads engineer, I've found that few people understand or appreciate the process of deriving structural loads for aerospace vehicles. A loads engineer is responsible for determining the forces, moments, pressures, etc., used to size aerospace structures. In essence, we must figure out how strong to build the vehicle and under what conditions it is permitted to break.

A common misconception is that the vehicle only needs to be built to withstand the nominal design mission, i.e., transporting people or payloads from Point A to Point B and repeating. In reality, a loads engineer must understand the vehicle and all of its systems well enough to consider and combine worst-case scenarios and failure modes, also known as off-nominals, that are likely to occur in the vehicle's lifetime.

For this reason, aerospace loads is as much an art as it is a science. What are the loads? People often expect loads to be deterministic based on a set of inputs, but in reality, the answer usually is "it depends." My first manager taught me that "all loads are wrong, but some are useful." The objective of a loads engineer is to err on the conservative side of wrong when accuracy is not practical, but not so conservative that the vehicle can't fly. This is where the challenge lies.

What's the best part of doing what you do?

Aside from the unparalleled experience of going to space and becoming an astronaut, the best part of my career has been the opportunity to learn a multitude of things by working on diverse projects. I've worked on helicopters, tiltrotors, flight simulators, airplanes, and spacecraft. As an analysis engineer, I've gained exposure to nearly every component of the vehicles I've worked on, and as a result, I've collaborated with almost every engineering discipline. I've been involved in every phase of the aerospace design cycle, from conceptual design to detailed design, all the way to flight testing and operational testing as a Mission Specialist and astronaut! This broad experience has allowed me to transition into the business side of aerospace engineering, which has been equally rewarding.

What is a necessary evil in your industry?

In commercial human spaceflight, we often find ourselves reinventing the wheel—questioning and reassessing aerospace industry standards and best practices. This is because many of these standards and practices either don't apply or conflict with the highly optimized weight and performance targets of reusable spaceflight vehicles. Additionally, flight test schedules are notoriously unpredictable, making it challenging to maintain a healthy work-life balance at times.

Have you ever changed your line of work? If so, why and what was the change?

Yes, I've changed my line of work a few times, with some changes being more significant than others. I started off in the helicopter industry as an external loads engineer, then transitioned to being a flight simulator engineer for helicopters. Later, I returned to external loads but for spacecraft and airplanes. Most recently, I transitioned from external loads (as a manager) to commercial strategy and then again to program management and integration. Each transition was driven by my curiosity and desire to learn new things. I'm a systems thinker at heart, so the more diverse and unique topics I'm exposed to, the more inspired I am to innovate, create, and solve new problems. Sometimes, the transition was self-motivated because I couldn't see myself at the end of the path I was on. Other times, external factors came into play, such as being offered an incredible opportunity to design human spaceflight vehicles or even go to space! In these cases, I was content on the path I was on, but the reward far outweighed the risks. Over the years, I've developed a framework that allows me to take calculated risks. For instance, I only concern myself with the risks associated with irreversible decisions. For everything else, I'm usually willing to take a risk, because if you never try, you'll never know. As Wayne Gretzky said, "You miss 100% of the shots you don't take."

What purchase under $100 has improved your life: career or personal?

For me, it would be either my Audible subscription or my Kindle. Both have made it incredibly easy to consume information and learn new things.

For about two to three years, my daily commute was about 140 miles round trip, which equated to over 2 hours in the car every day. Instead of wasting those 550 hours per year, I decided to use that time productively by consuming as many nonfiction books as possible. This period of long commutes and listening to audiobooks laid the foundation for many of the successes and opportunities that followed. After reading "The Millionaire Mind," I was inspired to read the biographies of successful billionaires, reasoning that if I could internalize and emulate just 1% of a billionaire's success, that's at least $10,000,000 worth of knowledge gained. This led to my "Billionaire Book Series," featuring the biographies of Richard Branson, Elon Musk, Jeff Bezos, Steve Jobs, Warren Buffet, Charlie Munger, and Oprah Winfrey. #LeadersAreReaders

How has social media benefited or hindered your career?

Social media, for all its pros and cons, has been an excellent networking platform for me. It's an easy and entertaining way to stay up-to-date on what's happening in the commercial space industry. It has definitely opened up some interesting doors for me. For instance, I never imagined I would walk as a model in New York Fashion Week, but thanks to social media, I've found a tribe of like-minded ambitious friends who have opened my eyes to possibilities I never thought were attainable. Using social media to further my career and explore complementary/supplementary career paths has been really valuable. However, it's important to use social media and other public media responsibly, ensuring alignment and proactive communication with your primary employer. Because social media is such a powerful tool, a few missteps can jeopardize your job.

STEM PRO
LAUREN-ANN GRAHAM

LAUREN-ANN GRAHAM: MS Mechanical Engineer, BS Aerospace Engineering, Orion Test Engineer at NASA

IG: *@nasa.lag*

LAUREN-ANN GRAHAM is a mechanical test engineer on NASA's Orion Spacecraft which will be carrying humans back to the moon! From the Launch Control Center, she supported the Artemis 1 mission management team during launch operations. Lauren-Ann received her B.S. in Aerospace Engineering and M.S. in Mechanical Engineering from the Florida Institute of Technology. Her ultimate dream is to become an astronaut and step foot on the moon! This dream has led her to explore many forms of engineering such as jet propulsion, test, aerospace, mechanical, and space systems. She hopes to inspire others to chase their dreams and shoot for the stars! Her favorite quote comes from her dad, "You only fail when you quit."

STEM PROs Questions

Why do you do what you do? / Do you have a defining/ah-hah/ eureka moment where you knew what you wanted to do?

My ultimate goal is to become a NASA astronaut so I have taken my passions alongside my goal to determine my decisions. For years, I've harbored a deep passion for understanding the intricacies of how things are built and function. It was my father who nurtured these curiosities, imparting invaluable skills in car repair, basic construction, and the art of crafting whatever I could lay my hands on. Fast forward to high school, I joined the robotics team to learn how to build robots given specific requirements and task goals. From then on I knew engineering was the right direction for me given my ultimate dream. Now as a test engineer, I am able to apply my passion for space and understanding of how things work, all while learning the systems of the vehicle I hope to fly in one day.

What is something you wish you knew while in school?

Engineering school was extremely difficult for me at times. I wish I could go back and tell myself that GPA is not everything. I applied to experiences I did not believe I qualified for due to my grades, but received incredible positions given I did not have perfect grades. My college journey began with me facing a significant educational gap compared to my peers, a consequence of the limitations imposed by my high school education. However, I did not allow those limitations to stop me from chasing my dream. I continued to study harder, tutor more, and surrounded myself with students who were succeeding. Do not count yourself out and never give up if it is your passion!

What is something you wish you did differently when you first started working?

I am still early in my career, but I wish I had asked more questions when interviewing for specific roles. The job description may not address major details, so it is important to ask many questions. This could help you to avoid roles that sound great on paper but may not align with your lifestyle.

If someone wanted to do what you do, what's the best piece of advice you'd share with them?

Surround yourself with a group of individuals who love and support you no matter the situation you are walking through. My family and friends have carried me through some of my most difficult seasons. They believed in me and encouraged me to keep going even when I did not believe in myself. I would not be where I am today without the help of some of the most important people in my life.

If you could definitively answer one unanswered science question, what would it be and why? / What unanswered scientific question keeps you up at night?

Personally, due to family members who struggle with incurable diseases, I would love to know the answer to curing autoimmune diseases. Lupus specifically is a disease that is completely unpredictable and misunderstood. It has the ability to ravage any part of the body with no rhyme or reason. Even until recent, the disease itself was very unknown to the mass population.

What scientific discovery or event in history would you like to go back in time and witness?

It would be a dream come true to be able to witness the day that man first stepped foot on the moon. The entire world was watching and holding their breath when Neil took that first step. People all over the world crowded around televisions to witness it. That is one moment in history humanity felt incredibly united over a success that the human species as a whole accomplished.

What's a misconception about your line of work?

Many people ask me if they need to be an engineer or be good at math to work in the space industry. There are so many positions that do not require math, science, or engineering to be successful! Companies still need business leaders, communications majors, and so many more to make their mission successful.

What's the best part of doing what you do?

I have the opportunity to be a part of history! There will be days when I am sitting inside the Orion Crew Module and think to myself "I am testing the systems that will carry humans back to the moon." Knowing my work will have a permanent impact on the advancement of technology and space exploration is a dream come true!

What is a necessary evil in your industry?

The aerospace industry can become extremely costly. It can provoke questioning amongst the general public because of the sheer amount of money being spent to develop the technology. However, without the industry we wouldn't have everyday items such as velcro, water filters, cell phone cameras, and even satellite television.

Have you ever changed your line of work? If so, why and what was the change?

Again, I am fairly early into my career so I have not needed to change my line of work yet.

What purchase under $100 has improved your life: career or personal?

The item that has improved my life the most would have to be my Bible. It has been a source of strength and direction through every season of my life. Whenever I face uncertainty, I am immediately brought back to peace in knowing God has ordered every single one of my steps.

How has social media benefited or hindered your career?

Social media has only benefited my career thus far. It has connected me with a wide range of incredible people in varying STEM industries. The platform has also inspired me in moments where I was struggling privately with battles that come with life. So many of my followers have inspired me, and I thoroughly enjoy helping others find their way in this industry!

STEM PRO

KELLY KNIGHT

KELLY KNIGHT: PhD Student in Science Education Research, MS Forensic Science & Forensic Biology, BS Chemistry & Forensic Chemistry, Associate Professor and STEM Accelerator of Forensic Science

IG: *@kellythescientist*
TT: *@kellythescientist*
X: *@ScientistKK*
FB: *Kelly the Scientist*
LI: *@kellythescientist*

KELLY KNIGHT, a 2020 University Teaching Excellence Award winner, is an associate professor of Forensic Science and a STEM Accelerator. As a STEM Accelerator, she mentors STEM students and leads K-12 STEM outreach programs.

After obtaining degrees in chemistry and forensic science, Prof. Knight worked in forensic science laboratories for almost a decade before starting her career in higher education. She has qualified as an expert in both forensic serology and forensic DNA analysis and has testified in several court trials.

In addition to her experience with forensic casework, she has many years of experience in research which has included areas such as laser microdissection and low copy number (LCN) DNA methods.

Professor Knight is also currently working on her PhD in science education research. Her research examines how out-of-school STEM programs impact BIPOC girls, particularly in the area of STEM identity.

STEM PROs Questions

Why do you do what you do? / Do you have a defining/ah-hah/ eureka moment where you knew what you wanted to do?

I am a forensic scientist and I do what I do because I truly enjoy my work. Not only is it interesting but it gives me a sense of satisfaction. Knowing I am contributing to something bigger than myself gives me the motivation to do my work.

I didn't have a single moment that led me to this career but instead I had smaller moments that all added up to me realizing this is what I wanted to do. There are two moments in particular that were especially salient though.

The first moment was a mock crime scene lab in my 11th grade anatomy and physiology class. This was the first time I experienced how all of the natural sciences I had been learning could be applied to a real-life career that had the potential to impact lives in the way that forensic science does.

The second moment was the internship I did my senior year because even though I was inspired by forensics in high school, it wasn't until I got to college and I decided what I didn't like that I was led back to forensics. I spent 3 years majoring in chemistry and then in my senior year I worked as a DNA technician in a DNA laboratory and having that firsthand experience really showed me that that was what I really wanted to do for the rest of my life.

What is something you wish you knew while in school?

I wish I knew how to better connect with mentors while I was an undergraduate. I truly felt lost in those four years. I struggled in my courses. None of my friends were majoring in STEM so I often felt alone. My professors felt distant and unapproachable. I wish I knew how to find my group of STEM folks who could have helped to provide me with more direction and encouraged me to do things I didn't know I should be doing like undergraduate research.

What is something you wish you did differently when you first started working?

Surprisingly, I do not have any regrets from when I first started working. When I was in graduate school, I finally got the mentor I needed and she really set me on the right path. I was particular when it came to choosing where I was going to work after graduation. I paid attention to the culture and the environment and asked questions during the interview that would help me to determine if it was a good fit for me. When I started working, I continued to network and meet people in the field and I took on opportunities that I thought would position me for growth in the laboratory.

If someone wanted to do what you do, what's the best piece of advice you'd share with them?

My advice is to read and learn as much as you can about forensic science from reputable sources. Follow real forensic scientists on social media and learn about their experiences. Ask if you can have a quick Zoom chat with them to learn more about the work they do. There are many different areas of forensic science such as crime scene investigation and DNA and the more you learn, the more you can decide which area interests you the most.

For those who are younger, I also suggest taking lots of science and math classes or joining STEM clubs at your school. These fields are fundamental to forensic science and will help teach you the important analytical and critical thinking skills you need to be a forensic scientist. I'd also recommend attending science fairs, STEM camps, or workshops to gain hands-on experience and exposure to different scientific fields.

For college students, I recommend seeking out undergraduate research opportunities and start looking today for ways to build the skills you need for the job you will want tomorrow. If your future career is going to be looking for experience with a particular type of lab technique or leadership skills or public speaking skills, you have to start think about what type of things you can get involved in while in college that will connect to those skills to be able to put on your resume or speak to in your cover letter.

And for everyone else, I can't express enough how important it is to find a mentor (or mentors) no matter what stage of your career you are in and to get connected with a group of like-minded individuals. Life is tough and we need other people around to help us navigate sometimes.

And lastly, always stay curious, and enjoy the journey of discovering your passion within the fascinating world of forensic science.

If you could definitively answer one unanswered science question, what would it be and why? / What unanswered scientific question keeps you up at night?

The unanswered questions I'd like to be able to definitively answer are pretty specific to my field of work, forensic DNA. Whenever we find a biological sample at a crime scene, as DNA analysts we try to individualize it (identify the source of the DNA) but even if we get a DNA profile and it is consistent with someone in the case (victim, suspect or other individual), there are a few questions we can't answer. We can't answer how long it has been there (time since deposition). We also can't tell you how it got there. In other words, is this a primary transfer of DNA or a secondary transfer (transferred through an intermediary from a different source)? These are questions that are often asked of us and that are important forensically but scientifically, we just aren't able to answer those questions yet.

What scientific discovery or event in history would you like to go back in time and witness?

I would have loved to have been there when Rosalind Franklin discovered the structure of DNA!

What's a misconception about your line of work?

This is definitely an easy one. This biggest misconception about my line of work is that it's "just like CSI". Many people are fascinated by forensic science because of all of the true crime documentaries and crime scene shows like CSI and NCIS but it is not like what they see on TV.

What's the best part of doing what you do?

Every day feels like a new puzzle to solve! I have been working in the field for almost 20 years and it still gets me excited every day. Now that I am a professor, the best part of what I do is knowing that I get to play a small part in training the next generation of forensic scientists to do this great work.

What is a necessary evil in your industry?

I would not necessarily refer to it as "evil" but a lot of folks in the field are not a fan of having to testify. It is an important requirement of our job but the idea of your work being questioned and scrutinized can bring on a sense of uneasiness for many forensic scientists. It is important that this scrutiny occurs though because even as forensic scientists, we are human and we need to be held accountable for the work that we do.

Have you ever changed your line of work? If so, why and what was the change?

I have stayed in forensic science since 2005 but I changed from being a full-time forensic scientist in a laboratory to becoming a professor. I changed because I was having my first son and my husband and I wanted to move closer to family. I was looking for jobs and had not considered a position in academia but when I saw this faculty role open, I applied for it because I had been doing some adjunct teaching while I was working in the lab and really enjoyed it. It was one of the best decisions I have ever made. I miss working in the lab but as a mom, the schedule flexibility I have as a professor is unmatched.

What purchase under $100 has improved your life: career or personal?

Honestly, a planner. I have so many moving parts going on in my life that I would be drowning without my planner!

How has social media benefited or hindered your career?

I would say that, overall, social media has benefited my career. I never intended to build a brand or become an "influencer" when I started my Instagram page. I started my page because I love what I do and I wanted to share more about science and my experiences in academia all while being a mom, but honestly, as much as I love my friends and family on my personal page, they weren't really interested in those topics. I wanted to have a space where I could connect with other people who had similar interests and now it has grown to be so much more than I ever dreamed. Science and academia are often thought of as rigid and serious but I wanted to show another side of it. It has been an amazing creative outlet for me. I also hope to inspire other women who are looking to get into these fields or are currently in these fields and are looking to connect with someone who can relate to their journey. These connections I have made have benefitted me tremendously because I am constantly inspired by my community and all of the amazing things they are doing. I've also been able to use social media to explore "STEMpreneurship" and turn my passions into profit by connecting with like-minded brands and organizations to amplify my mission and provide consulting services.

“The best way to predict the future is to invent it.”

–Alan Kay, computer scientist and winner of the A.M. Turing Award for his contributions to object-oriented programming languages and personal computing

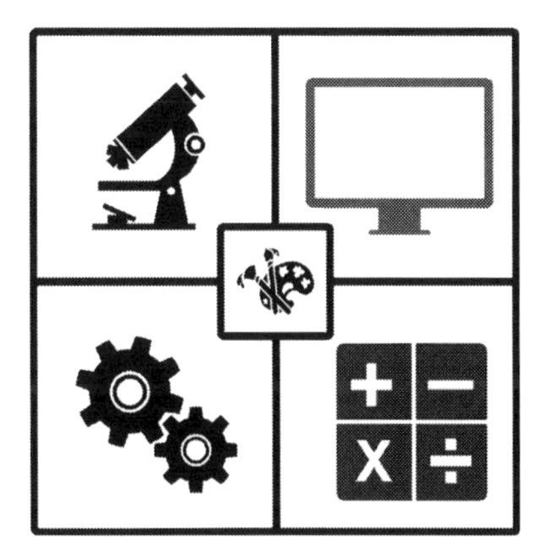

STEM PRO

JACQUELINE MEANS

JACQUELINE MEANS: Founder of the Girls Empowerment STEM Initiative; College student double majoring in Management Information Systems and Marketing with a minor in Neuroscience

IG: *@STEMQueenDE*
TT: *@STEMQueenDE*
X: *@STEMQueenDE*
FB: *STEM Queen DE*
YT: *@STEMQueenDE*
Website: *STEM-Queen.com*

JACQUELINE MEANS: Wilmington, especially Southbridge, is often associated with negative things. Shootings and drug deals are both things one may think of when someone says Wilmington, but Jacqueline Means is changing that. Jacqueline is a Management Information Systems and Marketing double major and Neuroscience minor at the University of Delaware. She is also the proud founder of the Girls Empowerment STEM Initiative, an organization she created 9 years ago at just 12 years old that is dedicated to bringing science, technology, engineering, and math, or STEM, to the underprivileged girls of Wilmington, Delaware. Jacqueline is certainly changing the stigma surrounding Wilmington by being an advocate for STEM education and volunteering her time to engage the youth of Wilmington and empower them to pursue careers in STEM fields; She does this by hosting Girls Empowerment STEM Events, workshops where she leads hundreds of girls in fun, hands-on science experiments! She has received many awards for her dedication to her community, as well as been featured on numerous national television shows, including The Today Show, The View, The Steve Harvey Show, The Kelly Clarkson Show, CBS's Mission Unstoppable (where she is a series regular), Access Hollywood, and many others. Learn more about her at www.STEM-Queen.com!

STEM PROs Questions

Why do you do what you do? / Do you have a defining/ah-hah/ eureka moment where you knew what you wanted to do?

Honestly, not super sure! I'm not 100% positive about what I want my career to be, but I know for certain that I want it to involve helping people.

What is something you wish you knew while in school?

That it's okay to say no. I was someone who always said yes to everything that was offered to me, which led to me being a part of just about everything, including Marksmanship, Chess Club, Business Professionals of America, Student Government, Cheerleading, National Honor Society, Dance, and more. In addition to all that, I of course had to stay on top of all my school work and my volunteer work, all of which meant I had a plate that was beyond full. Because I said yes to everything and took on way too much, my sleeping habits and mental health were incredibly low, and I felt exhausted every day. Looking back, I wish I had said no to more things. I feel that if I had, I would've been more able to pick and enjoy the few things I really did want to do, and I would have prioritized my sleep and mental health.

If someone wanted to do what you do, what's the best piece of advice you'd share with them?

Get started! It's never too early or too late to start giving back, so make today your day one.

What's the best part of doing what you do?

Engaging with hundreds of kids and seeing their faces light up as I teach them my experiments! It's so wonderful to see their eyes spark with amazement and delight, it is the absolute BEST part of what I do.

Have you ever changed your line of work? If so, why and what was the change?

I'm a full-time student, so not so much my line of work, but I have changed my major. I was originally a PreMed major, but decided to switch to Management Information SYstems and Marketing. My top two reasons for the switch were 1) I realized that I knew I wanted to help people, but I wasn't 100% sure that I wanted to be a physician, and 2) I love talking to and connecting with people, and I felt that a career in business would allow me to do more of that in my day-to-day work.

What purchase under $100 has improved your life: career or personal?

Honestly, a social media-blocking app. I did not realize how much of my day I was wasting on social media. I would spend hours and hours on my phone, not even realizing how much time was passing. Using the app has helped me spend less time on my phone and more time advancing my academic career and enjoying my life.

How has social media benefited or hindered your career?

I feel like this is in complete contrast to my previous answer LOL, but it has truly benefited my career. Social media in itself is a wonderful tool for connection. I am able to share the work that I do with not only my immediate community, but thousands of people from around the globe. Additionally, I'm able to meet amazing people that I would not have met otherwise, case in point: Kevin, the author of this book!

STEM PRO

ALLISON CUSICK

ALLISON CUSICK: Biologist, Oceanographer, Adventurer, Antarctic Specialist, PhD in Biological Oceanography (2024), MSc Marine Biology (2021), MAS Marine Biodiversity and Conservation (2017), BSc Biology (2006)

IG: *@womanscientist, @fjordphyto*
Threads: *@womanscientist, @fjordphyto*
TT: *@womanscientist, @fjordphyto*
X: *@woman_scientist, @fjordphyto*
FB: *Woman Scientist, Fjordphyto*
YT: *@womanscientist, @fjordphyto*
LI: *@womanscientist*
Websites: *womanscientist.com, fjordphyto.org*

ALLISON CUSICK is a Biologist, Oceanographer, and an Adventurer. She specializes in understanding the Antarctic ecosystem starting at the first layer of life, with phytoplankton in the ocean.

She is currently working at Scripps Institution of Oceanography, University of California San Diego, in the USA. Her first expedition to Antarctica occurred in 2013 where she lived aboard the U.S. Antarctic Programs' icebreaker Nathaniel B. Palmer for 53-days tracing the fate of algal carbon export in the Ross Sea. The morning of the day that she boarded the Palmer icebreaker for the expedition, she first ran a marathon on the Ross Ice Shelf, dressed as a banana.

Every year since 2017 (except the pandemic 2020-2021) she has been fortunate to travel to the Antarctic Peninsula on various tour ships running FjordPhyto as Lecturer, Scientist, and Expedition Guide. She has spent more than 300 days in Antarctica and hopes to continue. Allison has taken her work Internationally to Egypt talking about the importance of outreach and science communication at the United Nations Framework Convention on Climate Change (UNFCCC) COP27. Allison also had a brush with fame starring on the Disney + Jeff Goldblum show 'The World According to Jeff Goldblum: Tiny Things' episode. Recently, she earned her the Inaugural UCSD Chancellor's Innovation Award for her work in building up the FjordPhyto program a participatory science platform that intertwines climate science and polar expedition tourism fostering social engagement and ocean literacy.

STEM PROs Questions

Why do you do what you do? / Do you have a defining/ah-hah/ eureka moment where you knew what you wanted to do?

I decided to dedicate my scientific career to understanding polar ecosystems because I love learning about nature and how our world works. I love adventure and traveling to the furthest destinations humans can explore. I love working in the harsh environment in Antarctica and being reminded that earth is better when we embrace diversity and not myopic anthropocentric views. I love being mentally, emotionally, spiritually and physically challenged to grow. I want to inspire people to learn the curiosities of our world and how science works and be part of the solution for a healthy earth and social innovation. I have had a couple defining "eureka" moments in my life, but it took a while to experience them! In high school I thought I'd be an astronaut because they traveled to the moon and I loved to travel and explore. A lot of astronauts have Bachelor's degrees in STEM so I majored in Biology and Geology, because I couldn't pick just one! At that time I was accepted into the University of Washington's Husky Marching Band and while playing music I would have these moments where all of our sounds would transport me to a flow state. I know literally what it feels like to "be on the same wavelength" with others. In college I was sitting in Biology class and as the teacher explained plant photosynthesis and growth, I felt a similar 'wavelength' vibe, or flutter, in my heart thinking, "this stuff is SO cool! I need to know more" I now know that quiet internal feeling as "passion". I had a hard time after college really figuring out "what I wanted to be when I grew up". I was the first in my family to go to college, something I wasn't quite sure I would even do. No one in my family is a scientist, so I had no idea what it meant to be a scientist. To keep it short here, I worked for ten years doing different jobs after earning my B.S. Some felt sort of right, but not totally right. I tried a lot of things! I wrote a lot about this if you would like to visit www.womanscientist.com and read some musings of my earlier career journey! It wasn't until I had a serendipitous opportunity to work on an icebreaker in Antarctica in 2013, that I knew I wanted to dedicate my life to polar studies.

At age 30, I stepped off the C-130 Hercules plane and looked at the towering volcanoes and mountains and felt the numbingly dry biting air and thought, I feel like an astronaut who just landed on another planet—but I'm here, right on Earth! It wasn't until three years later, in 2016 that I frustratingly realized (after not being competitive in the job market with a Bachelors) that the only way to advance in science and pursue the dreams I had was to go into graduate school. At age 33 I started an Interdisciplinary Master's program in Marine Biodiversity and Conservation at Scripps Institution of Oceanography, UC San Diego (USA) because I wanted to know how to work my science background into societal change for policy, economics, and social issues in marine science. That program launched me on a path to create and grow a citizen science (participatory science) program called FjordPhyto where we partner with tour expedition vessels in Antarctica to help collect data and samples that we can analyze to understand how melting glaciers are influencing the first level of life—the phytoplankton in the ocean. It's been amazing to tap into the creative space of science and really dream up something and watch it come to life. And to see and hear how it's changing the lives of so many participants, travelers, students, and more! I hit the age of 40 and am just completing 6.5 years of my PhD work. Just to say, age should not be a limitation (young or old) to do amazing things!

What is something you wish you knew while in school?

In high school I wish I knew that you could create your own job. That you didn't have to just pick what was available. I wish I had realized that NO ONE truly knows the answers, everyone is just figuring it out. I spent a lot of time wondering who I am, what I wanted to do in life, what my purpose should be. I thought it would just be handed to me. I wish I had known earlier, that the creative spirit lives inside and the more you tap into what your true passions are, the more you can unlock it!

What is something you wish you did differently when you first started working?

I don't know if I could have known to do anything differently at that time in my life! I think a lot of the growth process is trying things out and seeing if it's right for you. For me, I wasn't quite sure of who I was and what I wanted out of life, and what I wanted to give back to society. I wish I had been more confident to move on when I knew things weren't right for me anymore. I had the bad habit of feeling guilty if I left bad jobs, or bad bosses, or bad situations. Like I was somehow a quitter and didn't tough it out. Or would be seen as unreliable and flaky. So, that is something I wish I had done differently, left bad situations and jobs earlier and not felt bad about it! Who knows how much sooner I would have had more eureka moments! We get ONE life, gravitate toward the things that make you feel alive inside, carpe diem!

If someone wanted to do what you do, what's the best piece of advice you'd share with them?

Have patience, resilience, and keep looking for mentors and people who inspire you! AND BE NICE TO PEOPLE! You never know when you will interact with them again and that is part of building a network, it's a slow process of being helpful and friendly and treating others with respect. For working in Antarctica I would say think of translatable skills. For instance, if you speak other languages—that could be a really convenient skill where everyone is so international. If you want to work on expedition ships, get yourself some water experience boating, being outdoors, wilderness first aid, build up your science communication and lecture skills, your people skills. These skills all build on themselves and it will take a while to build them all up. For resilience, don't let rejections get you down for too long. Rejection never feels good, but pick yourself back up and try again. Know that trying again means you're brave and courageous because most people don't even try. I applied to a lot of things, a lot of jobs, a lot of funding opportunities, and I got rejected from a lot of things. I've learned that it doesn't mean it's a NO FOREVER. It just means that wasn't the right opportunity, or it's a no for right now, or not the right timing. Don't give up! Always be ready to pivot.

If you could definitively answer one unanswered science question, what would it be and why? / What unanswered scientific question keeps you up at night?

I would love to know the answer to collective consciousness and this idea of kismet (destiny/fate), being on the same wavelength with people, "the world works in mysterious ways" things of the Cosmos—we have no way to scientifically answer those questions. I worked for 6.5 years as a technician at the Institute for Systems Biology and thought a lot about how individual parts make up a system that produces phenomena. I like understanding how the bigger phenomena come to be based on all the individual contributing parts of a system.

What scientific discovery or event in history would you like to go back in time and witness?

Because I think a lot about phytoplankton and photosynthesis, I would like to have been there when the first single celled organism bacteria, cyanobacteria, figured out how to use carbon dioxide to split water and make energy. We have evidence in the fossil record that occurred in Archaean rocks of western Australia, dated 3.5 BILLION years ago. That event changed the course of life on our planet.

What's a misconception about your line of work?

I'm not sure what all the misconceptions of my line of work are. Maybe that you can't work at sea if you get seasick? I get seasick!! Just take the medication. Another misconception might be that it's all fun and games and adventure. I know I make it look fun and adventurous on my social media, but that is also a slight misconception. I don't talk about the stress, the emotional toll, the time away from family and loved ones, nor the unfortunate occurrences of bullying and harassment, nor do I talk about the frustrating moments, the financial ruin it has taken to make it all happen. There are sacrifices made but the joys and positives pursuing my passions outweighs the negative. I think one misconception is that people may think it was all easy and I just waltzed through open doors. Some were open to me, for sure—for instance in Antarctica women used to be banned from working on the ice. Women weren't 'fit' enough physically or emotionally. Despite applying, they were never chosen for expeditions. Caroline Mikkelsen was the first female to set foot in 1935, the first USA female scientists to work at the South Pole were 7 women led by Dr. Lois Jones in 1969. Because of their persistence and spirit of adventure, the doors for other women to come down opened. I have found doors, banged on doors, tried a lot of doors, and created my own doors to open for others to walk through!

What's the best part of doing what you do?

I love what I'm doing now—being a scientist, working in the lab on genetics methods, working in the field, traveling on ships, thinking about ecological phenomena. I love that I have been able to use my skills and passions to teach in Antarctica (I've now spent over 300 days working on ships there!). I love getting others to feel excited about these places and about the microscopic world. I love that my career allows me to be flexible with when and where I work (no 9-5 job here!) gives me the creative and collaborative freedom working with awesome Colleagues, literally brainstorming solutions to answer some of the biggest unknown questions of our time! It's all very exciting.

What is a necessary evil in your industry?

To be honest, the line of work I've experienced there's never enough money. Environmental sciences don't get the same big bucks human health or for-profit initiatives get. It leads to a system of exploitation, relying on volunteers, offering low pay, strange power structures on who can write grants, or ask for funding. I don't think it has to be necessary but it's the system that was set up. Although I built a large portion of my resume/CV on volunteering, it contributes to the disparity in diversity in STEM. I want to change that by securing funding for people I collaborate with or try to make it a fair exchange in some way. I want to also get away from this mindset that somehow if you work in environmental science or not-for-profit you have to scrape by, pinch pennies, and stretch tight budgets. I want to be applying to large monetary grants to really build a vision and make immensely innovative things happen!

Have you ever changed your line of work? If so, why and what was the change?

YES so many times. I've never seemed to choose jobs that pay well, but I've stitched together multiple sources of income to make a greater whole. But I have always stayed in the theme of science. I had a 6-month internship in neuroscience at the Allen Institute for Brain Science helping to build the Mouse Brain Atlas, a precursor to the Human Brain Atlas. I then worked 3 years as a lab tech and mouse husbandrist in an Immunology lab. I realized I didn't like the scientific topics we were studying and quit, thinking I didn't like science. I was unemployed for two months looking for new jobs. Then I worked as a field biologist on seasonal jobs studying squirrels, songbirds, macaws. I loved it but it was minimum wage, seasonal work, or volunteer. I went back into a job in a lab at the Institute for Systems Biology studying the effects of ocean acidification on diatoms for 6.5 years. That job opened the door (serendipitously) to my experiences in Antarctica and oceanography. It was amazing but I also worked as a bartender on nights and weekends for three years because my non-profit lab job didn't pay enough. I then went into graduate school and during the first 3 years of my PhD (before the pandemic) I also worked as a technician 10 hours a week as a lab technician at Synthetic Genomics (now Viridos) genetically engineering algae for algal biofuel. I built my skills as a maritime professional and polar guide so that I could also start making income working as a lecturer and polar guide on expedition vessels 2-4 months every year (which conveniently also allowed me access to Antarctica as a scientist doing my PhD work on Antarctica while continuing to grow the FjordPhyto citizen science program) In doing all of these jobs and gigs and professions, I feel I have gained a larger perspective on many facets that people face. I value the changes I've gone through, always learning something new.

What purchase under $100 has improved your life: career or personal?

My Classpass monthly gym membership is under $100. When I discovered this app before the pandemic, it was a life changer. It allows me access to so many gyms and studios around my city and country. It's absolutely been a game changer as I cannot afford normal gym membership prices as a student and I can't be tied to one location due to my inconsistent travel schedule. The flexibility in choosing classes has allowed me to channel my stress and anxiety from career and personal challenges into healthy body positive outlets and allow me to train for the endurance, ultra, and tri races I so love feeling strong in doing.

How has social media benefited or hindered your career?

I have seen SO MANY BENEFITS from social media use. First, I do think everyone benefits from a little social validation and affirmation every now and then. So when I need a pick-me-up, or to be entertained or humored I find good funny uplifting content others create and share. I also have made so many connections with people who I sometimes have the fortune of meeting in real life, locally and around the world. I also like to use it as an opportunity to share my work and to also humanize myself as a scientist, to be personable and show my goofy, ambitious yet equally vulnerable personality with others. I have not yet experienced any hindrances in my career for using social media. Of course, you get comments from people assuming that if you're using it, you must not be working. We should all know by now that social media is curated and the appearance of "always being on social media" does not mean you're not doing other important things in life. I see it as a positive and a benefit, I use it to feel inspired. If I start scrolling and find myself feeling grumpy—I get off and go do something that energizes me—like getting out in nature or spending time in person with friends.

STEM PRO
YASMIN DICKINSON

YASMIN DICKINSON: PhD Student in Cardiovascular Medicine, Science Communicator, MS Molecular Biology, BS Biomedical Sciences

IG: *@the.cardiac.scientist*
TT: *@the.cardiac.scientist*
X: *@YasDickinson*
Website: *thecardiacscientist.com*

YASMIN DICKINSON is a Cardiovascular Medicine PhD student at the William Harvey Research Institute in London. Yasmin is working closely with AstraZeneca, focusing on the role of CNP (C-type natriuretic peptide) in signaling mechanisms involved with cardiac hypertrophy and heart failure.

Before diving into her latest degree (she also has a bachelor's in Biomedical Sciences and a master's in Molecular Biology), she worked as a research assistant at the University of Oxford, still in the cardiovascular arena, exploring hypertension in pregnancy and dealing directly with patient samples from women with preeclampsia.

STEM PROs Questions

Why do you do what you do? / Do you have a defining/ah-hah/ eureka moment where you knew what you wanted to do?

I have always been enthralled by science ever since I was young, but I remember being particularly interested in the cardiovascular system during secondary school. Not only is the heart and vascular system so intrinsic and beautiful, but there is so much more that we need to understand and uncover, in terms of pathology and treatment. I am the only person in my family to go into the STEM field, so before, I didn't really know what career paths there were in science, other than medicine. However, during my biomedical science undergraduate degree I undertook my own project for my thesis and fell in love with research, it was from this point onwards I knew I wanted to be a scientist.

What is something you wish you knew while in school?

I definitely wish I knew all the many amazing STEM fields available for little science-enthusiastic Yasmin. Students nowadays have loads of resources to learn about this sort of stuff, to feel inspired and to know what career paths are out there for them. But for me, I didn't have the faintest idea and my school did not give me much guidance either. All I knew was that I loved science and wanted to do good in the world using it. But hey, I still made it though!

What is something you wish you did differently when you first started working?

I wish I didn't spend so much time feeling down about making mistakes. Making silly mistakes in the lab is bound to happen, and in a way, it's crucial because we learn from them. Sometimes, we feel pressured to be flawless, to excel in every technique, to have all the answers, but that's not realistic. It's all a learning process.

If someone wanted to do what you do, what's the best piece of advice you'd share with them?

Please don't wait for opportunities to come to you, go out there and create the opportunities for yourself. This is a lesson I've learned firsthand and it's been one of the most valuable in my life. If you want lab experience, reach out to labs and ask to shadow scientists. If your university offers a mentorship scheme, sign up and speak with your mentor about internship opportunities. Also, spend time developing yourself in other areas. For instance, I knew I wanted to improve my communication skills and explore my creative side, so I took it upon myself to start my own science communication page on Instagram and worked on my written and spoken science communication skills. Since starting my page, I have learned so much about what I am capable of and what I truly love doing, and amazingly so many doors opened up for me by doing this.

What unanswered scientific question keeps you up at night?

I guess I wonder if we will ever truly understand human consciousness on a deeper level. Human consciousness is probably one of the most profound mysteries of the human experience, as it encompasses our awareness of life, thoughts, emotions, and our sense of self. On this note, we also do not know much about what actually happens during a near-death experience. People who have come close to death typically report vivid sensations such as out-of-body experiences, feelings of euphoria, seeing a white light or deceased loved ones. There is still so much we don't know! Sorry if this was a bit morbid haha.

What scientific discovery or event in history would you like to go back in time and witness?

Oooo I'm sure many would say this but probably the moment humans first set foot on the Moon in 1969. It was a monumental achievement and witnessing such a significant milestone in human history would have been unforgettable. A close second would be seeing the very first clear image of our planet Earth (like the Blue Marble photograph of Earth taken in 1972) for the first time!

What's a misconception about your line of work?

My field of work is within translational medicine and a common misconception is probably that scientists "can't be doing very much", since scientific discoveries, advancements and the development of new medications all take a very long time. This would possibly not be the case if there was more transparency on how research actually works. By understanding the reasons behind the time-consuming nature of research, such as conducting repeated experiments, the rigorous process of publishing, the many steps in clinical trials, and recognising their essential roles, people may begin to see why scientific advancements require patience and thoroughness.

What's the best part of doing what you do?

It's definitely the variety the role of a scientist entails! You're not just stuck in the lab, you become a public speaker when you present your work at conferences. You become a teacher when you partake in public engagement and outreach activities with school kids and the general public. You become a graphic designer when you have to create your own diagrams, schematics and images to explain your research. You become a writer when you write papers and contribute to scientific literature. Being a scientist has expanded my skillset in so many departments and I never get bored of my job.

What is a necessary evil in your industry?

A necessary evil in scientific research and academia is possibly the competitive nature of funding and recognition. While "competition" in academia can fuel a scientist's drive for innovation, it also creates a lot of pressure on researchers to constantly secure funding, publish frequently, and achieve significant breakthroughs to be able to advance in their careers. This sort of working environment can lead to stress and burnout, which is why joining the right lab is so important.

Have you ever changed your line of work? If so, why and what was the change?

I am probably in the midst of changing my line of work. I love being in the lab and conducting experiments, but through my science communication endeavors I've discovered so much more about myself and how much genuine joy I get from communicating science with others and doing so through creating videos and teaching in person. I am still doing my PhD but maybe I'd like to continue down the sci-comm route after I finish, time will tell!

What purchase under $100 has improved your life: career or personal?

Probably my Fitbit—I really needed it to monitor my sleep—and make sure I get enough of it!

How has social media benefited or hindered your career?

When I first started my science communication page on Instagram, I would have never guessed just how many wonderful opportunities and friendships I'd make along the way. I originally started it as a hobby and a means to hone my creative side, but it evolved into so much more, allowing me to network with other scientists and work with big brands such as Thermo Fisher Scientific and the TV show CBS Mission Unstoppable. I was also interviewed for a leading magazine in my country Bahrain which was a meaningful milestone for me. I also gained more confidence in myself through pushing myself to try new things, which encouraged me to lead STEM events such as Pint of Science and New Scientist Live. However, the most rewarding aspect of this all has been the opportunity to meet so many incredible scientists and creators through social media. We constantly inspire one another, and being part of this supportive network has been truly invaluable.

STEM PRO
ELIO MORILLO

ELIO MORILLO: MEng Systems Engineering & Design, BS Mechanical Engineering, Blue Origin, formerly NASA JPL

IG: *@thespacemechanic*
Threads: *@thespacemechanic*
FB: *The Space Mechanic*
TT: *@thespacemechanic*
YT: *@thespacemechanic*
Website: *thespacemechanic.com*

ELIO MORILLO'S fast-paced career arcs from Ecuadorian, Puerto Rican, and New Yorker roots to influential space systems engineer and esteemed author, captivating audiences with his dedication to space exploration. A master's graduate from the University of Michigan, Elio's work melds mechanical, electrical, and systems engineering expertise, contributing to breakthroughs like NASA-JPL's Mars Perseverance rover, Ingenuity helicopter, and Blue Origin's lunar MK1 lander. His memoir, "The Boy Who Reached for The Stars," not only traces his professional journey but also underscores education's transformative power, the immigrant experience, and the pursuit of space for all. Elio's keynotes and mentorship inspire many, exemplifying his commitment to expanding human potential beyond our world.

STEM PROs Questions

Why do you do what you do? Do you have a defining/ah-hah/ eureka moment where you knew what you wanted to do?

I believe humanity's destiny lies among the stars. The universe's vast expanse and its untapped resources promise a future where Earth can flourish even more. The journey begins with perfecting technology for Earth-lunar transport. Learning from our moon, our closest celestial neighbor, will pave the way for humanity's leap further into the solar system.

My passion for aerospace technology was ignited early on, fueled by stories of the monumental efforts that culminated in lunar landings. However, understanding the darker chapters of space exploration, including the 20th-century space race and Operation Paperclip's aftermath, underscored the importance of pursuing future endeavors with a strong ethical compass. It is crucial that we remember the lessons of the past and strive for equity in our quest to explore the cosmos, for the benefit of Earth and all its inhabitants.

What is something you wish you knew while in school?

While technical expertise is indispensable, emotional intelligence plays a crucial role in navigating the complexities of professional environments. The essence of effective leadership lies in emotional intelligence. Effective communication and teamwork often present greater challenges than the technical problems themselves.

Understanding and prioritizing wellness—physical, emotional, mental, and spiritual—is fundamental. No one else will prioritize your well-being for you.

What is something you wish you did differently when you first started working?

I wish I had understood the importance of setting boundaries earlier in my career. Diving into work with little regard for personal well-being led to feelings of inadequacy and underappreciation, despite significant efforts. It became clear that hard work alone does not guarantee recognition or success. Establishing boundaries is essential for maintaining health, motivation, and relationships within the workplace.

If someone wanted to do what you do, what's the best piece of advice you'd share with them?

Build stuff. Whether it's coding, building robots, or designing complex systems, hands-on experience is invaluable. Technical knowledge is crucial, but the skills developed through teamwork and collaboration are irreplaceable. Engage in projects that challenge you and offer opportunities to work with others, fostering a deep understanding of both the technical and interpersonal aspects of engineering.

If you could definitively answer one unanswered science question, what would it be and why?

The existence of life beyond Earth fascinates me. Confirming the presence of extraterrestrial life would not only be a monumental discovery in the scientific community but also redefine humanity's understanding of its place in the universe. It's a question that bridges scientific curiosity with profound philosophical implications about our existence and the possibility of life in the cosmos.

What scientific discovery or event in history would you like to go back in time and witness?

Witnessing the pivotal moments when ancient civilizations first mapped the cosmos would be awe-inspiring. These early astronomers laid the groundwork for our understanding of the universe, combining observation, mathematics, and philosophy. To be present as they charted stars and planets, using the sky as a canvas for their intellect and imagination, would be an unparalleled experience.

What's a misconception about your line of work?

Many assume that working in spacecraft engineering means I'm an astronaut. While the prospect of space travel excites me, and I do hope it becomes a reality in my lifetime, my role focuses on designing, building, and testing spacecraft. It's a common misunderstanding but one that highlights the diverse range of careers and contributions within the field of space exploration.

What's the best part of doing what you do?

The opportunity to contribute to our understanding of the universe and to push the boundaries of human capability is incredibly rewarding. Being part of a team that designs and tests spacecraft, determining their capabilities and limitations, is a constant source of inspiration and challenge.

What is a necessary evil in your industry?

The notion of a "necessary evil" does not align with my perspective. All choices, especially those involving ethical considerations, are just that—choices. The challenges we face should be approached with integrity and a commitment to positive outcomes. Dismissing any aspect of our work as a "necessary evil" undermines the potential for innovation and ethical progress.

Have you ever changed your line of work? If so, why and what was the change?

While I have remained dedicated to spacecraft engineering, I am always open to exploring new avenues that enhance education and foster the next generation of engineers and scientists. My commitment to education and mentorship is a testament to the belief that sharing knowledge and inspiring others is as important as the technical achievements we strive for.

What purchase under $100 has improved your life: career or personal?

Several modest investments have significantly enhanced my daily life, both personally and professionally. A high-quality travel pillow, an online course on ice plunges, a bidet (I can't highlight this enough, especially to Americans), and subscriptions to various apps that streamline productivity and encourage digital detoxing have all contributed to a more balanced, healthy, and efficient lifestyle.

How has social media benefited or hindered your career?

Social media has been instrumental in sharing my journey and inspiring others to pursue careers in STEM. It has enabled me to connect with a global audience, sharing insights and experiences that demystify space engineering and highlight the importance of diversity and inclusion in our field. While it presents challenges in maintaining privacy and managing time, the benefits of outreach and community building have been invaluable. This includes opportunities to write my memoir, be a keynote speaker, be on TV, and meet some incredible people who have become close friends!

STEM PRO

DR. SERGIO SANDOVAL

DR. SERGIO SANDOVAL: PhD Aerospace Engineering, Guidance Engineer at NASA

IG: *@sergioa_sandoval*
LI: *Sergio Alfonso Sandoval Escobedo*
Email: *info.sergiosandoval@gmail.com*

SERGIO SANDOVAL is a Mexican-American aerospace engineer at NASA's Johnson Space Center. His area of expertise is in entry, descent, and landing and he works as a guidance engineer developing the guidance commands that will go onboard the vehicles that will land humans on the Moon with Artemis. He earned his PhD in aerospace engineering, where he focused on the optimization of entry and powered descent trajectories for future astronaut missions to the Moon and Mars. As part of his doctoral research, he has developed the first powered descent guidance abort plan since the Apollo program. He is passionate about outreach and has given close to 100 presentations across the United States, Mexico, and Latin America where he shares the motivational story about his beginnings coming from Mexico without knowing English and following his dream to become an aerospace engineer.

STEM PROs Questions

Why do you do what you do? / Do you have a defining/ah-hah/ eureka moment where you knew what you wanted to do?

I work as a guidance engineer because I love the impact it has in the world. Everything we do is connected in some way and when you find your passion, it is a lot easier to feel that you are part of something. Working for NASA, we impact everybody, so I feel like I am part of Earth, not just a specific place.

What is something you wish you knew while in school?

I studied aerospace engineering, and the focus was always on that. I wish that I had more opportunities to learn about other things in life such as philosophy, psychology, finances, etc. I had some of these classes, but they were always applied to specific areas of business, never on a personal level. I had to find those resources on my own to realize that the world is not just full of engineers. We are all part of a big system where everybody's talents are important.

What is something you wish you did differently when you first started working?

I wish I would have been more honest with myself about the things that I actually know. I didn't want to let anybody down, so I would give myself additional tasks to learn skills that I didn't have just to do a specific task. Over time, I learned to be honest to myself and say that I needed time to learn or that there might be somebody better for that specific task. At the end, it didn't affect me or my job and it took away extra effort and stress that could have been avoided.

If someone wanted to do what you do, what's the best piece of advice you'd share with them?

This is going to be a marathon and the hardest thing you'll ever do in your life. The only way you are going to make it through is by remaining humble and keeping supportive people with similar goals throughout the journey.

If you could definitively answer one unanswered science question, what would it be and why? / What unanswered scientific question keeps you up at night?

I am a big fan of Einstein's work and I always felt incomplete by the unfinished "Unified Field Theory." The idea that there is a "Theory of Everything" that would link general relativity and quantum mechanics sounds fascinating. It is a lot easier when answers make sense to us and we can find order in things, and not having this solution might be the reminder we need that we don't know everything. I would love to know if there is an actual link, and you can define both big and small universes with the same theory.

What scientific discovery or event in history would you like to go back in time and witness?

I would like to have been around when Orville and Wilbur Wright made their first flight in 1903. When I started as an aerospace engineer, I wanted to make airplanes. One of my earliest undergraduate memories was learning about the Wright brothers and the first flight in Introduction to Aerospace Engineering. In a way, it is thanks to them that we are able to go to space now.

What's a misconception about your line of work?

That we never landed on the Moon, but that one is too simple to prove. A more personal one is that it is common for some people to think that everybody that works at NASA is an astronaut. There are a myriad of people working in many different missions, some of them not even involving astronauts, to make sure that every single one of them is a success. Although many of us would like to go to space, our role is to plan, develop, and execute each mission from beginning to end as engineers, scientists, or technicians.

What's the best part of doing what you do?

I get to help humanity in everything I do. It might be hard to see this because many times it is under several layers, but the truth is that working at a place like NASA, I get to serve the people of the world in many areas such as education, outreach, exploration, commerce, research, among others. The cool work and missions to space are just the tip of the iceberg, every single launch and mission has a purpose that will ultimately benefit all of us on Earth.

What is a necessary evil in your industry?

Fuel consumption and waste through disposable spacecraft. This is the biggest environmental issue with the space industry, you can't really get around the usage of fuel. We are slowly moving towards reusable components, but even then, not everything can be reused. The impact of space exploration far outweighs these downfalls, but hopefully in the near future we find ways towards cleaner and more efficient solutions.

Have you ever changed your line of work? If so, why and what was the change?

Back in high school, I thought I was going to study marketing. I am a first-generation college student, and I didn't have a lot of guidance on how to choose a major. Ultimately, my friends thought that I would be a much better engineer than a marketer (and I believe that is completely true. When I found out about engineering, I never looked back, it has been my passion ever since. Along the way, I did some work in the business and education world, and although I loved them, I always went back to engineering, particularly, aerospace engineering.

What purchase under $100 has improved your life: career or personal?

Under $100 you can buy a lot of books that can change your life, so I will list three of my favorite ones:

- *The Girl With the Dragon Tattoo*: Back when I was 17, I learned to speak English with this book. I didn't gain any lessons, it was just a fun book, but I read word by word with a dictionary on the side to learn a new language. It changed my life because I realized that I could achieve any goal without having anything but motivation.
- *The 5 People You Meet in Heaven*: This book can change your life, I read this book over and over because it ground me to reality and the importance of the people around you.
- *The Richest Man in Babylon*: One of the most challenging parts of being a PhD student is learning to survive with your student stipend. This very small book literally put my life in order.

Honorary mentions:

- Calm/Headspace: I have been meditating for over 5 years and I know it has improved my quality of life greatly. Anything that affects your mental health positively is worth having.
- Things: This helped me move from countless notes, post-its and papers all over the place, to one single place where all my to-do thoughts live now.

How has social media benefited or hindered your career?

I am neutral to that. I like to think that I can make a small impact on people that haven't been able to meet me in person or that can't easily reach me. But to me, it doesn't add anything to my career. I love sharing my story out of my own personal time and money, sometimes going as far as paying the trips out of pocket just because I know how valuable outreach is. But social media is not who I am, and it has always been hard to share my lifestyle on a platform that is not part of my lifestyle. Building on that point, time is the most valuable thing that I have, and social media has sometimes taken away from it for no apparent gain.

STEM PRO

EMILY P. SETO

EMILY SETO: Manager of Planetary Protection, Contamination Control and Research Scientist at Honeybee Robotics, formerly NASA

IG: *@mightymicrobe*

EMILY P. SETO is a Planetary Protection and Contamination Control Engineer. She has supported notable flight projects such as Mars2020, Mars Sample Return, Dragonfly as well as international space projects such as JAXA's Martian Moons Exploration (MMX). Serving as the principal investigator for numerous NASA research grants, she maintains active collaborations with leading research institutions and industry partners. Presently, she spearheads lunar biology In-Situ Resource Utilization research, with the goal of facilitating human presence on the lunar surface.

STEM PROs Questions

Why do you do what you do? / Do you have a defining/ah-hah/ eureka moment where you knew what you wanted to do?

The practice of Planetary Protection entails protecting solar system bodies from contamination by Earth life and protecting Earth from possible life forms that may be returned from other solar system bodies. Essentially, it revolves around the principle of ensuring that microbes on Earth do not inadvertently hitch a ride to other planetary bodies and conversely, that we do not introduce contamination to our own biosphere. As a citizen of Earth, I believe we should do our due diligence and prevent contamination on other planetary bodies. Just as we have learned valuable lessons from dealing with invasive species on our own planet, we must apply these insights to our interplanetary activities. Introducing microorganisms to extraterrestrial environments could disrupt delicate ecological balances, hinder scientific investigations, and potentially compromise the search for indigenous life beyond Earth.

What is something you wish you knew while in school?

Growing up in a traditional household, I was instilled with the belief that only a handful of career paths were deemed worthy of pursuit. From an early age, the mantra echoed: become a doctor, a dentist, a pharmacist, or something similarly esteemed—further exacerbating the pressure to conform to conventional professions. It was during moments of introspection and self-discovery that I realized success is not confined to a predetermined set of professions. Rather, it is shaped by passion.

What is something you wish you did differently when you first started working?

For two years, I juggled three part-time jobs. The first job stretched from 6:00 AM to 2:30 PM, the second from 3:00 PM to 11:00 PM, and I taught piano throughout weekends. Often, I found myself catching sleep in my car between shifts, leading to an unhealthy lifestyle. While this hustle helped chip away at my student loans and provided clarity on my career path, looking back, I would have advised my past self that it's acceptable to take breaks, reduce shifts, and prioritize my health.

If someone wanted to do what you do, what's the best piece of advice you'd share with them?

During my time at NASA, I collaborated with engineers to support flight missions such as Mars 2020, Europa Clipper and Mars Sample Return. Despite lacking an engineering background, I proactively immersed myself in the engineering domain, adopting a humble mindset and avidly absorbing information. By actively engaging with engineers, familiarizing myself with research and taking on additional responsibilities, I overcame challenges in a field unfamiliar to me. Currently, at Honeybee Robotics, I continue to grow and seize opportunities in leadership by leading projects, writing, securing NASA research grants and leading a Planetary Protection and Contamination Control team. Most recently, I started biology in-situ resource utilization research, aiming to support humans on the lunar surface. Being proactive is key!

Have you ever changed your line of work? If so, why and what was the change?

I navigated a non-traditional career path, transitioning from the medical field to the aerospace industry. Initially, I found myself immersed in the clinical laboratory setting, conducting tests and contributing to patient diagnoses. However, as I delved deeper into my work, I became fascinated with research and made the decision to pursue a Master's degree in Clinical Microbiology. Throughout my graduate studies, I focused my project on spore-forming organisms. It was only through research that I realized specific species of spore-forming organisms were also found in NASA cleanrooms. I was fascinated by how the microorganisms in cleanroom environments adapt and develop resistance to specific sterilization techniques and space conditions.

STEM PRO

DR. YASMINE DANIELS

DR. YASMINE DANIELS: PhD in Analytical Chemistry, Certified Industrial Hygienist in the US Government, Adjunct Professor, Author, mentor, mom

IG: *@classychemist*
Threads: *@classychemist*
TT: *@classycheminist*
X: *@classycheminist*
FB: *Classy Cheminist*
YT: *@classychemist*

YASMINE DANIELS has a Ph.D. in Analytical Chemistry with a focus on Environmental Remediation and has conducted research aimed at developing fundamental tools to help design eco-friendly ways of removing toxic substances (like heavy metals) from the environment. With over 20 international and peer-reviewed journal publications, she is well-known within the science and social media community as the "Classy Chemist." Dr. Daniels is also an Adjunct Chemistry Professor, an Occupational Safety and Health professional and STEAM advocate. She currently works full time as a Certified Industrial Hygienist in the US government, reviewing and assessing industrial & chemical hazards. She is also a full time mom and wife, volleyball coach and mentor.

In 2021, she wrote a bestselling children's book, *Building My Self-eSTEAM in Science* which helps to motivate youth to pursue STEAM. Her book has since appeared on Amazon Bestselling lists in three individual STEAM categories:

1. Children's Engineering Books
2. Children's Computer Hardware and Robotics Books
3. Children's Math Fiction.

In 2023, she wrote an award-winning children's book, *Black and Brown are Beautiful Crayons too!*, which was featured on ABC7 News, on a segment which highlighted books that inspired and uplifted children during a time when literacy rates had plummeted.

Dr. Daniels has been highlighted as a Chemist who is, "breaking barriers" by the American Chemical Society Axial Journal and as a "Chemist Star" by the Chemical and Engineering News Magazine. She has a passion for ensuring that students are provided the tools that foster an equitable learning environment and she hopes to continue to inspire and support youth through her work.

STEM PROs Questions

Why do you do what you do? / Do you have a defining/ah-hah/ eureka moment where you knew what you wanted to do?

I'm always torn about how to answer the question about why I do what I do because I do SO much. I'm never sure which answer to give. If I were to choose one of the things that I do, and then answer the question of why I do it, I would choose teaching chemistry. I'm truly passionate about teaching chemistry because I want to ensure that other students get to experience the nurturing and support that I had while I navigated chemistry courses as a student myself. Chemistry is not the easiest subject so having a professor who can enable you to not only understand, but to fall in love with the subject is absolutely rewarding.

What is something you wish you knew while in school?

I wish I knew more about the very professions that were available within the field of chemistry. I didn't learn about them until well after I graduated and having an idea of those fields would probably have pushed me a bit harder. As a chemistry student, I was only aware of about four fields that I could enter into. Those being medicine, big Pharma, academia and government.

If someone wanted to do what you do, what's the best piece of advice you'd share with them?

My best piece of advice is to never close yourself off to an opportunity. In fact I advise folks to cast a wide net. In doing so, you open yourself up to opportunities that may challenge you but that will eventually reward you as well. Oftentimes we limit our capacity to what we think we should do, when in fact, there are unique opportunities that we may not be exposed to until we create the space to be exposed to them.

What's a misconception about your line of work?

People often think that I am a dental assistant or dental hygienist when I tell them I am a Certified Industrial Hygienist. I don't blame them though—it's definitely an underrepresented profession that I wish more people knew about. Industrial Hygiene is the science of protecting the health and safety of people in the workplace by anticipating, identifying, evaluating and controlling chemical, physical and biological workplace hazards. It's the reason that many people get to go home safely to their families every day.

What's the best part of doing what you do?

The best part of doing what I do is knowing that I am helping to save lives. In addition to that I know that through my work, I continue to educate, whether it's educating students, families, workers or organizations.

Have you ever changed your line of work? If so, why and what was the change?

I used to work full time as a research chemist. Although I am grateful for that experience, I knew that I wanted to experience something different. For a long time, it was just me and the test tubes and my chemicals. I'd wake up everyday, go to the lab, run my experiments and then churn out data and publications. Occasionally, there would be a conference or two where I'd present my research but for the most part, my public interaction was limited. I decided to transition into industrial hygiene because the profession allowed me to continue to apply the science that I had already known since IHs (industrial hygienists) use lots of analytical portable tools to take samples and measurements of chemical and other hazards. As an analytical chemist that made me feel right at home. On top of that, I got to speak to and educate workers, employers, unions, and other professionals in the workplace almost daily. Becoming an industrial hygienist has been one of the best decisions that I've made in my career.

What purchase under $100 has improved your life: career or personal?

A journal. I have more journals than I care to disclose, ha! Honestly, as wild as my imagination can be sometimes, it helps to be able to capture my thoughts and to organize them in a journal. I have one for just about every category of idea that I am thinking.

How has social media benefited or hindered your career?

Social media hasn't really impacted my professional career (i.e., my 9-5 as an IH) but it's certainly done wonders for my entrepreneurial side. The side where I conduct STEM outreach, publish children's books, mentor youth, etc. With that said, there are some things I wish social media wouldn't do. Although it's a great place to connect, network, advertise and showcase your work or talents, it's also a breeding ground for idea theft, creativity infringement and unoriginality. I say that because I take so much pride in the things that I do and the time that I take to present it to a public audience and it's really a shame when I see other social media "influencers" recreate these original or unique posts, sometimes verbatim. You've gotta have tough skin on social media. You just have to know how to pick yourself up. Eventually the authenticity in your work will speak for itself and it will be the loudest voice in the room—or on that social media platform.

"We have the opportunity to create the future and decide what that's like."

—Mae Jemison, former NASA astronaut, engineer, physician, and the first Black woman to travel into space

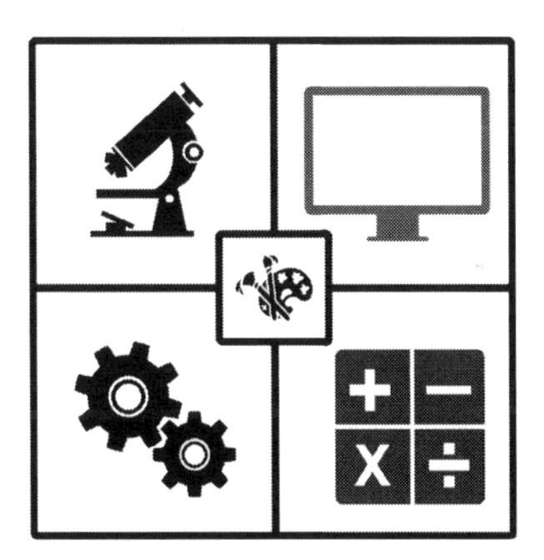

STEM PRO
DR. SHANNON COFIELD

DR. SHANNON COFIELD: Ph.D. Geological Oceanography, U.S. Navy Veteran

IG: *@geo_shanz*
X: *@MarsRoverMapper*

SHANNON COFIELD is a Geological Oceanographer with the U.S. Bureau of Ocean Energy Management (BOEM) where she works with coastal sediment resources, as well as exploring for deep sea critical minerals. She earned her Ph.D. in geological oceanography from Old Dominion University, Norfolk, VA. She's also a classic "non-traditional" student as she served in the U.S. Navy with the Presidential Honor Guard before transitioning back to college. Her focus on geology and sedimentology paved the way for multiple types of research. Her Ph.D. focused on deep sea Arctic sedimentology, and in 2016, she got a chance to apply her geology knowledge to another planet. Shannon interned with NASA's Jet Propulsion Laboratory and joined the NASA Mars Science Laboratory Curiosity Rover Science Team to develop geologic maps of Gale Crater, Mars. Following her internship, she was selected as a NASA Virginia Space Grant Fellow and helped map Jezero crater, Mars, which was ultimately selected as the NASA Mars2020 Perseverance rover landing site.

Her passion for exploring extreme environments is only surpassed by her love of teaching and STEM outreach.

STEM PROs Questions

The views expressed here are my own and do not reflect the views of my employer.

Why do you do what you do? / Do you have a defining/ah-hah/ eureka moment where you knew what you wanted to do?

I was always interested in nature and STEM. Some of my classic childhood stories are asking my parents how birds could perch on power wires or being equally terrified and fascinated of tornadoes. I always remember seeing museum exhibits of researchers bundled in puffy red coats in Antarctica, or inside vehicles in the depths of the ocean, and wondering how I could "be like that'.

My path was not a straight shot, but rather a windy mountain trail. I enjoyed my STEM classes in high school and attempted to make the straight jump to college. I was a little lost though. I couldn't decide which path to choose and I had this unrealistic, self-imposed pressure that I had to make the decision during my freshman year. Students, please don't put that pressure on yourselves! Enjoy college and keep an open mind in those general education courses... they might open some unexpected doors. While I was window shopping courses, 9/11 occurred and changed my entire trajectory.

I served in the U.S. Navy for four years and intended to use my military benefits to finish college and return as a Naval Officer. I landed in a Navy town, Norfolk, Virginia, and started my oceanography journey at Old Dominion University. I went through the process naturally by exploring my interests and following new doors. I quite literally looked through the university course catalog and flagged things that interested me. Everything fell into place.

Since oceanography is interdisciplinary, I didn't have to immediately choose my focus (biological, chemical, physical, or... geological!). I took one geology general education course and realized, "That's it—I'm going to study rocks (eh hem, sediments, underwater volcanoes, tsunamis, mountains, and so much more)!" Then, I discovered graduate school, and that's where the real magic happened! In graduate school, you do a deep dive into your field, start a research project, and slowly become more independent—which means more freedom to follow your own ideas, questions, and hypotheses... enter the Ph.D. Again, all of this was a natural evolution, as I didn't have my eyes set on graduate school during undergraduate.

Fast forward to the steep hill on my windy mountain trail—finishing my Ph.D. Not only is that journey a bit of a rollercoaster, but I was also trying to make big decisions about my career. I love teaching, so I originally followed the academic path of a post-doc position, then joining an academic department. One of my fellow graduate students was working for a small team with the federal government which was tasked with managing all federal marine minerals. I had some preconceived notions of federal positions from my time with the U.S. Navy, but this team was quite different. They covered a wide variety of duties (do you see a trend in my preferences, yet?) including regulatory, working with other federal agencies on research projects, developing and overseeing research projects, and regular stakeholder engagement for coastal—and—deep sea marine minerals. A rare position opened for a geologist and oceanographer, and the rest is history! If there's a secret, it would be to follow your passions, keep an open mind for new tangents, and don't think you need to choose one path.

What is something you wish you knew while in school?

During my undergraduate studies, I wish I had awareness of the Research Experience for Undergraduates (REU) program. It's a nationwide program that pairs undergraduates with academics from a variety of fields to gain experience in the lab, field, and working on graduate-level research projects. During graduate school, I had the opportunity to serve as the REU Coordinator for my department, so I witnessed our students rapidly grow and become more confident in their STEM skills. As a humble brag for their exceptional abilities, I recently attended one student's conference presentation on her dissertation research, and another student works for my sister federal agency!

If someone wanted to do what you do, what's the best piece of advice you'd share with them?

First and foremost, I would encourage everyone to follow their passions. If you were interested in becoming an oceanographer and/ or a geologist, there are plenty of ways to explore this field from grade school through early college (yes, even as a citizen scientist)! In grade school, let your mind explore and be curious. Allow yourself to go down information tangents, get lost in encyclopedias, and wander around museum exhibits. There's wonderful content on multiple online platforms (Kevin is an excellent resource!).

If you are in high school, don't think you need to make that big career decision. It can guide you through your next steps—and you can be laser focused, but it's not mandatory. Keep following your interests, polish up those math, communication, and STEM skills. I know math gets a mixed rating, but I promise it becomes fascinating as you progress through calculus. That spiral shape of a galaxy or the way a succulent grows—guess what—it's math (Fibonacci, to be exact). Never underestimate your abilities. We all did it one day at a time.

When you're ready for college, check out what certain departments offer. Do they have a wide variety of STEM programs, or do they specialize in one field? Do they have any partnerships with other universities or industry? For example, Old Dominion University is located in a major trifecta of military bases, NASA, and the Jefferson Lab (a particle accelerator facility). My university is also adjacent to the Virginia Institute of Marine Sciences and William and Mary, so students have various opportunities to participate in research projects or collaborate with all these facilities and organizations. For graduate school, reach out to Researchers with interesting projects. Ask them about their research or if they'd be interested in accepting a new graduate student. You never know what doors may open until you knock.

If you could definitively answer one unanswered science question, what would it be and why? / What unanswered scientific question keeps you up at night?

Let's be honest, selecting one question is nearly impossible. People in STEM have curious minds and we are always pondering mind-bending questions. I share the company of many researchers when I ponder, "What existed before our Universe developed?", or "What is the singularity in a black hole (or what is dark matter? Seriously, what is it!?!). As a geologist, I think on vastly different timescales than most researchers. We think in billions or trillions of years, so I'd like to know what Earth and our Solar System will be like in, say, the next 1 billion years. Along the same vein, geologic time is the Forgotten Player when thinking about life on other planets. When we see light from other galaxies, we are looking back in time. What if that snapshot of time is the equivalent to Earth's life 3 billion years ago? We wouldn't see much, but if we look at Earth (or those distant planets) today, it's an entirely different story. Not only does that distant light need to reach Earth but it also has to be in the perfect geologic time that life existed on that distant planet. Those are pretty tough statistics!

What scientific discovery or event in history would you like to go back in time and witness?

I'm going to step outside the box and say I'd love to see what the moon looked like when it was still going through its volcanism and cooling stages. Granted, standing on the Earth during that time period wouldn't be exactly practical since Earth is likely still cooling, too... and no free oxygen courtesy of cyanobacteria yet. I imagine it would be a spectacular sight to see our moon glowing during its initial formation.

What's a misconception about your line of work?

Hands down, the most common misconception is that geological oceanographers are the same as marine biologists; therefore, "we must study dolphins". Don't get me wrong, dolphins are fascinating and extremely intelligent, but that's not quite the focus of this field. Geological oceanographers study rocks, sediments, minerals, plate tectonics, seafloor characteristics, coastal processes, marine geohazards (such as tsunamis), water quality as a factor of sediment concentration, how ocean currents impact sediment movement, interactions between marine life and geology, the deepest trenches on Earth, underwater volcanoes, black smokers with unique chemical properties, and so much more! Just as it sounds, it's quite interdisciplinary. While we don't study dolphins, the type of geology in their environment does impact their available food sources and maybe even the games they play!

What's the best part of doing what you do?

I truly enjoy the interdisciplinary nature of my field. If you ask most scientists, we could hold a number of titles. Broadly, I'm a Geological Oceanographer, but if I was asked for a more specific title, I'm a Sedimentologist and a Geochemist. Both of those suggest a plethora of interdisciplinary work. Sedimentology requires the "break down" of rocks, which includes physical and chemical processes in marine and terrestrial environments, and atmosphere. Rocks on a mountain may start to erode through rain and snow, fall down a slope, take a trip down a river to the ocean, get stuck in a coastal environment as waves push the sediment onshore and offshore, some of the grains may get blown around by the sea breeze, and they may eventually end up buried in the ocean sediments. What a journey! Sedimentologists sample those regions and analyze sediment's geology and chemistry ("geochemistry"). We act as detectives and piece together a grain's entire journey including where it came from. We use those same techniques on other planets and planetary bodies! However, if we didn't look at the entire picture—physical, chemical, biological, and geological—we would miss some vital clues in our sleuthing.

What is a necessary evil in your industry?

For me, the most challenging—and important—skill is communicating. There are different levels and types of communications, and they usually resonate differently to everyone. While I enjoy STEM outreach, the content tends to be more focused or packaged in a "Goldilocks" volume (i.e., not too long, not too short, not too complex, not too simple). Those interactions also tend to have momentum from the beginning as they hitch a ride on current events, such as an upcoming launch or research announcement.

However, there's another type of communication that tends to be a bit stressful. It's the base of the Scientific Method = reporting your findings. Sounds easy, right? Just write about what you did in your research or experiment? For any given experiment, there are massive amounts of content. Trying to piece together years of methods, results, statistics, discussion, and ALL those figures... it gets overwhelming quickly. During my dissertation, I ran into this "information overload" challenge frequently. Sometimes it causes you to freeze. If you're reading this and going through these challenges, please know it's okay and normal.

I always like to offer some solution when presenting a challenge. First and foremost, find a team. That team can be other graduate students, online academic communities, or anyone supportive to your style of productivity. Second, take time on the weekend to relax. I know how difficult that can be given your stress levels, but your mind needs time to decompress and recharge. Third, start by writing your methods, or a portion of the material you're comfortable with. It will serve as a confidence booster, give you some momentum, and, hey, you have to do it anyway! My final secret for those really tough moments. In a thesis or dissertation, we have a section devoted to thanking those who helped us through the journey. Take a second to think about that section or revisit it. My entire family, university department, and all my friends supported me throughout my graduate work and research. The best way I could convey my gratitude was by successful finishing and defending my research.

Hopefully some of those suggestions will help. When in doubt, a good cup of coffee or tea and walk in nature always helps! Just remember, one step at a time.

What purchase under $100 has improved your life: career or personal?

When I sat down in my first college Physics course, a fellow student recommended a handheld manual that's worth its weight in gold. I still have my original copy and it's my go to reference for everything from a list of Calculus sequences and series to thermodynamic equations. Highly recommend having a copy of Engineering Formulas by Kurt Gieck and Reiner Gieck (I have the 7th Edition) to assist you during your college coursework. It goes without saying, a geologist must always have an inexpensive set of loupes on hand, and a trusty rock hammer.

Personally, one purchase that improved my life would be the adoption fee for my sweet, scraggly rescue Fox Hound, Dash. Granted, she costs much more than $100 each year (she loves her treats and squeaky "buddies"), but she fills me with instant joy every day!

How has social media benefited or hindered your career?

While social media has not been a major focus of my career, it has definitely proven to be a useful tool. I have connected with a variety of researchers throughout the world and leveraged their knowledge for helping my own research challenges. There's power in having a platform where I can post a quick question and receive cited answers in less than a day. I have also utilized social media for collaborations and staying current on upcoming events, conferences, and research publications. I love a good "live Tweet" session from academic conferences!

Kevin has shown the true power of social media by inviting a diverse group of STEM backgrounds for his Takeover Tuesday project. I always enjoy meeting his other participants and learning about their passions and research.

STEM PRO
MORGAN DENMAN

MORGAN DENMAN: Space-focused Graphic Designer & Illustrator, NASA HQ

IG: *@morgandenman*
X: *@MDenmanDesign*
Website: *morgandenman.com*
Etsy: *Morgan Denman Design*

MORGAN DENMAN is a space-focused graphic designer and illustrator. She is passionate about combining art and science to effectively and interestingly communicate, educate, and inspire about space. Currently, she is a graphic designer for NASA Headquarters as a contractor, where she works with the Office of Communications. In addition to her day job, Morgan also freelances for other companies within the aerospace industry. She also makes time for personal design and illustration projects, which she sells as products in her Etsy shop.

STEM PROs Questions

Why do you do what you do? / Do you have a defining/ah-hah/ eureka moment where you knew what you wanted to do?

Art is something that I've always enjoyed. When I was little, I would spend hours each night filling my many sketchbooks before going to sleep. Every chance to draw, paint, and be creative was one that I took and enjoyed. I had no idea I could make a creative career a reality until my high school art teacher showed me what graphic design was. Once I discovered graphic design, I knew that was what I wanted to do with my life. Creating art with a message and a purpose was everything I wanted to do. It wasn't until after graduation from college that I discovered what would be the message and purpose behind my work. Working at my first job as a graphic designer at Kennedy Space Center Visitor Complex is where I discovered that message and purpose. I knew next to nothing about space when I began my job there, but the more I learned, the more I fell in love with it. It wasn't long after I started working there that I knew I never wanted to work in any other industry. Being a part of history in the making and being responsible for helping to educate and inspire the public about what's going on in the space industry is work I love and work I care very much about. I can't imagine doing anything else.

What is something you wish you knew while in school?

I wish I knew that NASA and the space industry had internships and jobs for more than just engineers. I always thought that you had to be a "rocket scientist" to work for NASA or any other space company, but in reality, many more career types contribute to the success of their missions. Creative and communications careers also play an important role in those successes. If I had known that, I might have applied for an internship at NASA or another company while in school and discovered my passion for space communications and design sooner.

What is something you wish you did differently when you first started working?

When I first started working, I prioritized work above all else. Work came before things it should not have, including my health. I thought that was what I was supposed to do to be successful and excel at my job. In doing this, I undervalued my time. Instead of completing my work to the best of my ability during set working hours and leaving it at that, I regularly worked far more than 40 hours a week for a job with a set salary. I wish I realized earlier that while it is good to work hard and try to hold yourself to a standard of excellence, it should never be at the expense of your life and your mental and physical well-being.

If someone wanted to do what you do, what's the best piece of advice you'd share with them?

The advice I would give to someone who wants to do what I do is to start creating. You don't need to have a client or a "real job" to be able to make work you are proud of. If you want to design mission patches, illustrate posters, or create artistic social media content, then do that. Design, illustrate, and create the work you dream of being paid to do, then share it. Put it out in the world on social media, your portfolio website, or wherever you can. Passion projects are an incredible way to showcase your skills and passion for a specific type of work/subject matter. By creating and sharing consistently in this way, you will build a great portfolio to show to the jobs/clients you dream of working for/with.

If you could definitively answer one unanswered science question, what would it be and why? / What unanswered scientific question keeps you up at night?

I want to know with certainty if we are alone in the Universe or if intelligent life exists beyond our fragile planet. There is so much left to discover about the Universe and our place in it. That definitive answer would change everything, hopefully for the better.

What scientific discovery or event in history would you like to go back in time and witness?

The Apollo 11 mission. I imagine watching the launch of a Saturn V rocket and witnessing humanity's first steps on the Moon, and the thought of that gives me goosebumps.

What's a misconception about your line of work?

A big misconception about my line of work is that it's easy. There's so much more to it than just making pretty space art. It takes time, skill, and practice to learn and understand the many design programs and the tools within them, especially since they are changing and evolving with new technologies. It takes extensive knowledge of color theory and the principles of design. It takes a thorough understanding of the subject matter you are creating art for. And it also takes a willingness to never stop working to grow and develop your knowledge and skills. All these things and more are what it takes to be a successful graphic designer. It's far from easy. It takes a lot of work. But all that work is worth it.

What's the best part of doing what you do?

I get to wake up every day and create art for a subject matter and a mission I am passionate about.

What is a necessary evil in your industry?

A necessary evil in my industry is feedback and criticism. As a graphic designer, I still identify as an artist. I put my all into my work and feel proud of the work I create. When I first started design courses in college, the criticism and feedback of peers cut like a knife because it felt personal. However, over time, I learned that feedback and criticism are essential to grow as a designer. My peers were not critical of my work because they did not like me or did not think I was a good designer. They were helping me to see that work differently. Their comments opened my eyes to new possibilities I might not have seen otherwise, and because of that, my work improved. In my career now, when I ask a creative peer or a client for feedback, I take what they have to say and look at it as an opportunity to further learn and grow as a designer and artist.

Have you ever changed your line of work? If so, why and what was the change?

I have always been a graphic designer and illustrator, but the areas I specialize in have changed, specifically in my freelance work. I used to take on almost any client or job when I first started, but now my freelance design clients are exclusively space-related. Even more specifically, what I specialize in is mission patch design. I love the symbolism and importance of the meaning of mission patches. I take pride in creating something representative of the collective work and mission of hundreds, if not thousands, of people. That is the reason why I chose to specialize in the design of mission patches.

What purchase under $100 has improved your life: career or personal?

There is not one single item that stands out, but one category that does. That category is books. I have books filled with design references and inspiration, books on creative thinking, books about space and space history, and many more. If I ever feel stuck creatively, opening a book and finding inspiration in one form or another often remedies that feeling.

How has social media benefited or hindered your career?

In large part, I credit the success I have had thus far in my career to social media. By sharing my space-focused passion projects on social media, I built a supportive network and became a part of the larger space community. The people I connected with on social media have shared my work and supported me. That support has led to so many incredible job and project opportunities I've been fortunate enough to have.

STEM PRO

KSENIA OZKOK

KSENIA OZKOK: Mentor, Astropreneur, Podcast Host, Space Content Creator, Founder @Re.Brand Academy, Co-Founder @BluDot App

IG: *@spacexenia*
LI: *@spacexenia*
YT: *@astropreneur*
Websites: *spacexenia.com, theastropreneur.com*

KSENIA OZKOK is an entrepreneur with over a decade of experience in marketing and communications. She has lent her expertise to renowned companies and organisations such as UNOOSA, SpaceWatch. Global, Siemens, OMV, and Spacebit. She founded Re.Brand Academy, an EdTech company dedicated to empowering professionals to enter the space industry. Beyond providing access, the academy offers guidance and mentorship focused on enhancing leadership skills, advancing careers, and building personal brands. This initiative is part of her mission to develop the workforce of the space industry by fostering tomorrow's role models. Ksenia was recognized as one of the "Top 100 Women in Aerospace and Aviation to Follow on LinkedIn." Additionally, she served as an Ambassador for the International Astronautical Congress (IAC) in 2023. She holds an MA in International Development and a BSc in Business and Economics.

STEM PROs Questions

Why do you do what you do? / Do you have a defining/ah-hah/ eureka moment where you knew what you wanted to do?

I've never dreamed about space. The only thing that connected me to space was the fact that I was born in Kazakhstan, home to the Baikonur Cosmodrome. Back then, I was working at a tech conglomerate. Despite its prestige, I never felt fulfilled, and the environment wasn't pleasant for me. At the same time, I was pursuing my MA in International Development, and the UN was always a hot topic. One day, I randomly decided to take a chance and submit my application to several job opportunities, based on my experience in Communications and Marketing, then hoped for the best. After three months, I had nearly forgotten about these applications. It was while clearing out my email inbox that I noticed an email from an "unrecognized sender": UNOOSA. "What is this?" I pondered, nearly dismissing the email as spam before something urged me to pause, noticing the "UN" in the sender's address. The email was an invitation to interview for an internship in Technical Applications, seeking someone with a background in communications. After interviewing, I was invited to the UN's Vienna head office, marking the start of my journey into space. Quitting my corporate job, I discovered a spark for space, inspired by people who were genuinely passionate about their work.

What is something you wish you knew while in school?

Dreams do come true, so don't be shy—dream big. Dreams are free, and limits exist only in your mind.

What is something you wish you did differently when you first started working?

I wish I had started working on my personal brand and content earlier and maintained more consistency with it.

If someone wanted to do what you do, what's the best piece of advice you'd share with them?

I would share my YSP method guide with them, so they could start building their brand more effectively. The YSP Framework offers a holistic approach to developing an authentic personal brand. It's designed to unlock your unique strengths, enhance your communication skills, ultimately improving leadership skills and boosting your career or business.

If you could definitively answer one unanswered science question, what would it be and why? / What unanswered scientific question keeps you up at night?

How did life originate on Earth?

What scientific discovery or event in history would you like to go back in time and witness?

Experiencing the space launch of Yuri Gagarin would indeed be a momentous occasion. To witness the vibe, the power, and the social cohesion during that time would be incredibly special. There is something truly magical about being part of a moment when humanity took a giant leap forward, united in awe and wonder at the possibilities that lay beyond our Earth.

What's a misconception about your line of work?

Most space professionals aspire to leave a legacy and make a significant impact on our society—and beyond—with their discoveries. They recognize the necessity of developing their expertise to achieve this. However, many overlook a critical element of the Impact formula: Visibility.

Expertise + Visibility = Impact

Consider this: imagine you possess the greatest masterpiece ever painted by Picasso, yet it's merely collecting dust in your attic.

Nobody will be able to see it.

Nobody will be able to witness its beauty.

Nobody will be able to appreciate the work of art you own.

Now, envision you dust off that painting and place it prominently in your living room. Suddenly, every person who catches a glimpse of this masterpiece can't help but stop and stare, left in awe by its beauty. But when the painting is hidden away with the Christmas decorations in a corner of your attic, no one can admire it. Why? Because they're not aware of its existence.

The same principle applies to you, my fellow readers. People can't see all your talents, even if you're an exceptional aerospace engineer, scientist, space lawyer, space marketer, business developer, etc. You might possess immense potential, but the public can't appreciate your professional skills, expertise, and knowledge unless you shine the spotlight on yourself. That's why becoming visible in the space community is so crucial.

What's the best part of doing what you do?

The best part of what I do is witnessing the tangible successes of those I've guided. Imagine someone who felt lost coming to me for help, and then, months later, they're making headlines, winning awards, securing leadership roles, attracting investments, smiling on billboards, and appearing on Forbes lists. This isn't something I have to imagine—it's a cherished reality and the most rewarding aspect of my work. It fills me with an immense sense of gratitude, knowing that my contributions have a meaningful impact.

What is a necessary evil in your industry?

A necessary evil in our industry is the neglect of personal branding. I've observed that many in the space community shy away from the limelight, hesitant to showcase their achievements, or leverage social platforms. There's a common misconception among these individuals that personal branding equates to mere "self-promotion"; It's about turning your volume up in a world that's too busy to listen. You've got to put yourself out there if you want your work to be visible. How are you supposed to make an impact if nobody knows you exist? The world needs to know your name, understand your work, and recognize your contribution. Only then can you truly make waves.

Have you ever changed your line of work? If so, why and what was the change?

Yes, I've experienced several shifts in my career trajectory, but the most interesting to speak about is the change within the space industry. After leaving my corporate position for the space sector, I was struck by the number of talented individuals around me. These people had immense expertise and skills, yet they often faced a lack of confidence, along with deficiencies in public speaking and communication skills, which in turn, hindered their ability to attract opportunities and achieve career growth. This realisation led me to shift my focus from a B2B model, where I provided marketing services to space companies, to a B2C approach, serving individuals directly. By doing so, I aimed to offer a mentoring program specifically tailored to build an authentic personal brand, alongside leadership and communication skills.

What purchase under $100 has improved your life: career or personal?

A customised phone case. Every year I set a new behaviour pattern that I strive to improve. Currently, the new pattern I'm willing to elaborate on is "consistency"; before, it was "speed". Since a phone is the attribute that I carry every day, most of the time, and if I don't carry it, it's still next to me. I decided to customise my phone case by engraving the aimed habit. Looking every day multiple times reminds me about my goal and, as a result, every time I was able to improve my behaviour tenfold.

How has social media benefited or hindered your career?

Social media has definitely benefited my career. I encourage everyone to harness its power because it's been instrumental in amplifying my impact. Daily, I empower hundreds with my content and through engaging conversations in chats, where I provide guidance on how to pursue a career in space, find their path, and stand out. The benefits extend significantly to career and business opportunities as well. For example, my Instagram played a crucial role in making me one of the influencers for the world's biggest international space congress, IAC 2023. In May 2023, I shared a series of Instagram stories about SpaceX attending the Congress, tagging the Space Agency of Azerbaijan, Azercosmos, who then noticed me. A month later, we met at the GLOC conference in Oslo, where I offered to interview their representative about the upcoming congress for my Astropreneur YouTube Channel. A few weeks later, they invited me to Baku to start a 100 days countdown to the congress. They were incredible hosts, providing me with a tour around Baku and its region to showcase all aspects for the congress participants. Following this, they invited me to attend the congress as a content creator. This experience opened up even more opportunities for me. Hence, when used wisely, social media can indeed serve as a launchpad for professional growth.

STEM PRO

PARSHATI PATEL, PH.D.

DR. PARSHATI PATEL: PhD in Astronomy/Astrophysics, Planetary Science and Exploration, Science Communicator and Educator, Author, and Artist

IG: *@parshatipatel*
TT: *@parshatipatel*
X: *@parshatipatel*
FB: *Spacegeek.parsh*
YT: *@parshatipatel*
Website: *parshatipatel.com*

DR. PARSHATI PATEL is an astrophysicist, science communicator, educator, author, and artist based in London, Ontario. Born and raised in India, Parshati moved to Canada to study astronomy. She has a Honours. B.Sc. in Physics and Astronomy from University of Toronto as well as a M.Sc. and a Ph.D. in Astronomy and Planetary Science & Exploration from Western University. During her graduate studies, she researched young, massive stars and the disks surrounding them using both observational data and computer modeling. During her Ph.D., Parshati developed a passion for STEM education and outreach while participating in several on-campus outreach programs. Naturally, post-Ph.D., she transitioned into the field of science communication.

For over a decade, Parshati has shared her passion for space, fostering awareness among the masses, particularly youth. She accomplishes this by spearheading award-winning space outreach initiatives, organizing 300+ educational public events, delivering 75+ invited talks and contributing regularly to the media both in-person and online with over 100+ interviews so far! Previously, she has worked at Western University's Institute for Earth and Space Exploration as Educational Outreach and Communications Specialist as well as on an assignment to the Canadian Space Agency as the Program Designer and Education Advisor in the Youth STEM Initiatives team. At the time of this writing, Parshati is a freelance professional bringing her expertise to a wide range of organizations and audiences.

In addition to consulting for science books, she has also authored a children's book called 'My Books of Stars and Planets' with DK Publishing. Parshati serves on the Board of Advisors for Students for Exploration and Development of Space (SEDS)—Canada and Zenith Canada Pathways. Parshati has a Professional Certificate in Communications and Public Relations as well as a Certificate in Digital Communications, Innovative Thinking and Project Management from Western Continuing Studies.

Parshati often shares her experience as a woman of colour venturing into the field of astronomy and barriers she had to overcome as an international student, with the hope of helping others. When she isn't communicating science, you will often find her in her art studio experimenting with fluid art techniques to create space and nature-inspired paintings. Alternatively, if the skies are clear, you will find her capturing the night sky through the lens of her cameras in the dark sky locations. Parshati likes to use her creative side to inspire people to be curious about our surroundings.

STEM PROs Questions

Why do you do what you do? / Do you have a defining/ah-hah/eureka moment where you knew what you wanted to do?

I was first introduced to what lies beyond the few stars and the Moon in the night sky during a chance visit to a planetarium when I was a teenager. The visit sparked my curiosity about the universe prompting me to delve into books to learn more about it. However, it wasn't until I received a gift of a telescope that I knew that I wanted to be an astrophysicist and study how the stars and planets form. The entire process of setting up the telescope and observing the craters on the Moon and Saturn's rings through my own telescope propelled my excitement about exploring further about the universe.

I proceeded to study astronomy and earned a Ph.D. During the process of obtaining my doctoral degree, I became passionate about bridging the gap between the research being conducted by astronomers and the public's understanding of space. I feel incredibly fortunate to be able to merge my technical knowledge and love for space with my passion to ignite the public's curiosity about space in my career as a Space Science Communicator! My role involves developing engaging space activities, workshops, and programs tailored for youth, educators, and the public. Additionally, I have the privilege of sharing my knowledge and enthusiasm for space through speaking engagements and media interviews.

What is something you wish you knew while in school?

While I was focused on my coursework and research during my time in undergraduate and graduate school, I now realize the significance of engaging in extracurricular activities. Looking back, I wish I had taken advantage of these opportunities to broaden my skill set. Participating in extracurriculars offers invaluable experiences that complement academic learning and provide practical skills applicable in the professional world. If I could go back, I would encourage myself to explore diverse activities and pursuits to enhance my abilities beyond the classroom.

What is something you wish you did differently when you first started working?

One thing I wish I had done differently when I first started working is maintaining a comprehensive record of my tasks and achievements. In retrospect, I realize the value of documenting everything I accomplished. There were instances where I performed tasks or acquired skills that I hadn't included in my resume or portfolio. Each task and experience contributed to my growth and knowledge in unexpected ways. Now, I prioritize logging everything I do to ensure I have a thorough list of achievements to draw upon when applying for jobs or showcasing my experience.

If someone wanted to do what you do, what's the best piece of advice you'd share with them?

The best piece of advice I would share with someone aspiring to pursue a career like mine is to follow your passion and embrace opportunities to step outside your comfort zone. When I was younger, I used to dread public speaking engagements. However, I recognized that overcoming this fear was essential for personal and professional growth. Instead of aiming to become a skilled speaker from the outset, I focused on conquering my fear and improving gradually. As a result, I now regularly give lectures and presentations as a science communicator. Embracing challenges and pushing beyond your comfort zone opens doors to unexpected opportunities and personal development.

If you could definitively answer one unanswered science question, what would it be and why? / What unanswered scientific question keeps you up at night?

The question of whether aliens exist is something that truly keeps me up at night. It taps into the fundamental curiosity of humanity about our place in the universe. As a researcher in star and planet formation, I've come to understand that the building blocks for life as we know it are prevalent throughout the cosmos. This leads to the interesting question: if the ingredients for life are abundant in the universe, could life like that found on Earth be common elsewhere? Moreover, I often ponder on the fact that we only have one example of life—life on Earth. This raises a profound question of whether our understanding of life is limited to what we've observed, or if there are entirely different forms of life beyond our current comprehension. Exploring the existence of aliens isn't just about satisfying curiosity; it could potentially revolutionize our understanding of the nature of life itself.

What scientific discovery or event in history would you like to go back in time and witness?

The scientific discovery I would like to go back in time and witness is when Edwin Hubble discovered the existence of other galaxies beyond our own Milky Way in 1923. Hubble identified a Cepheid variable star in the Andromeda Galaxy which he used to calculate its distance. His calculation showed that Andromeda was much farther away than previously believed, demonstrating that it was not a part of the Milky Way but an independent galaxy. This discovery revolutionized our understanding of the universe, leading to the realization that the universe is much larger and more diverse than previously thought. It has just been over 100 years since the discovery and since then we have learned so much about the universe.

What's a misconception about your line of work?

One common misconception about my field, astronomy, is its confusion with astrology. I often get asked if I can predict people's future since astrology is a practice that claims to provide information on human affairs and events by studying the positions of celestial objects. However, astronomy is a field of science where one studies objects as well as phenomena in the universe beyond the Earth's atmosphere. To avoid the confusion, I often specify that I am an astrophysicist! Astrophysics, closely related to astronomy, is a field of science that applies principles of physics to study astronomical objects and phenomena in the universe. For science communication, there's a lack of awareness about it being a viable career path. Many individuals don't realize the breadth of opportunities available in this field for those passionate about sharing scientific knowledge with the public.

What's the best part of doing what you do?

The best part of my work is undoubtedly bridging the gap between the technicalities of space and making it accessible to the public. What I find most fulfilling is engaging with youth, particularly when conducting activities and answering their questions.

Have you ever changed your line of work? If so, why and what was the change?

Yes, I've undergone a significant career change. While pursuing my Ph.D. in Astronomy, I initially planned to follow the traditional path of academia, aiming for a postdoctoral fellowship and eventually a professorship. However, upon completing my Ph.D., I reassessed my priorities. Moving to another country for a postdoc would mean uprooting my family, and I wanted to avoid that disruption. Additionally, during my doctoral studies, I developed a passion for space education.

Recognizing the opportunities in this field, I decided to transition from being a researcher to becoming a space educator and science communicator. It was a decision fueled by my desire to remain in Canada and to pursue work by combining my passion for space and education. This change allowed me to leverage my expertise and experience in astronomy while fulfilling my aspiration to engage with the public and inspire the next generation of space enthusiasts.

How has social media benefited or hindered your career?

Social media has been a tremendous asset in my career, enabling me to network within my field and draw inspiration from fellow science communicators. Through social media, I have had the opportunity to connect with professionals worldwide, learning about their impactful work and often collaborating with them or offering support. Without social media, I would have been restricted in the scope of individuals I could engage with within my field, highlighting the invaluable role it plays in fostering collaboration and innovation in science communication.

STEM PRO
SOPHIA CROWDER

SOPHIA CROWDER: Student Ambassador @HigherOrbits, NASA/Virginia Space Grant Consortium Scholar, Founder of IT girls & STEMSQUAD, 2023 Jefferson Scholar

IG: *@sophiacrowder_,*
@southsidestemsquad
FB: *Southside STEM Squad*
LI: *@Sophia Crowder*

SOPHIA CROWDER is a Student Ambassador for Higher Orbits and founder of ITgirls & STEMSquad—Sophia aspires to inspire the next generation of STEMists, Innovators, and Space Explorers. She is a staunch advocate for gender equity and founded an initiative (ITgirls) to help empower other young girls to pursue non-traditional career pathways such as those in IT & STEM. She is an active participant in speaking to a wide range of audiences from all over the world in webinars and online programs, live panels at conferences, and as a TedEd speaker. She also initiated a mentoring program in her community by leading STEMSquad, an educational after-school program for kids K-8th. Sophia serves in the Advisory Board of Être Girls and in the IfThenSheCan Girls Advisory Council (both a mentorship platform for female empowerment). In 2019, Sophia was part of a team whose STEM experiment with Higher Orbits won a ride to space to create art in microgravity. Aside from being a spokesperson for STEM and girl-empowerment, Sophia is also an avid musician (violinist with the Duke University String School); multi-awarded essayist (who has earned national merits in the Patriot's Pen & Voice Of Democracy competitions; as well as a Teen Correspondent/Author at Worth Magazine); she is also a roboticist & coder; trained martial artist (karate & fencing); as well as a certified scuba diver and pilot-in-training. She has an Associate's Degree with concentration in Science from Southside Virginia Community College (Class of 2023) and has been awarded the prestigious Jefferson scholarship as well as named a Rodman scholar, both merit-based, at the University of Virginia School of Engineering and Applied Science, where she hopes to pursue a double major in Engineering (Systems/Aerospace track), and Astronomy, as well as minor in Computer Science. Sophia has been selected to participate in various renowned programs such as the 4H Teen Excellence in Leadership Institute (TELI); Future Leaders in Algorithm, Mathematics, & Engineering Sciences (FLAMES) at NCSU; the International Science School in Australia (representing the USA); and the Engineering Camp with the Society of American Military Engineers (at USAFA & USNA). She was also a three-time scholar of the NASA/Virginia Space Grant Consortium selected to participate in highly competitive programs such as the Virginia Space Coast Scholars (VSCS), Virginia Aerospace Science & Technology Scholars(VASTS), and the Virginia Earth System Science Scholars (VESSS). Her biggest aspiration is to be an innovator for the space program, as an engineer and scientist. She is 18-years-old.

STEM PROs Questions

Why do you do what you do? / Do you have a defining/ah-hah/eureka moment where you knew what you wanted to do?

Growing up, I've always been passionate about pursuing whatever I am curious about—whether that be dinosaurs, building things, or space! But I didn't start out as a space nerd. When I was little I was actually a dinosaur-aficionado and I had all the toys and books to prove it. You can ask me what the differences between a Brontosaurus, a Diplodocus, and a Bracchiosaurus are and I can give you a detailed comparison like no other 4-year-old could. But it was that love for dinosaurs that propelled me to look up at the sky and wonder about all things space. I remember sitting in the kitchen one evening and asking my mom why there aren't living dinosaurs anymore and that's when she told me about how this fiery rock from the sky scorched the earth and killed the dinosaurs off. I think that was my "eureka moment" as it then led me to develop a curiosity for all things space. I wanted to find out how and where and when and why a ball of fire from the sky could wipe off the planet along with my favorite reptiles. From there, my interest in space science has only grown and ultimately became the object of both my curiosity and passion to this day. At the moment, I am studying to be a (Space) Systems Engineer with a double major in Astronomy.

What is something you wish you knew while in school?

I wish I knew that...it was okay to fail as long as you keep trying. A mentor once told me that (as cheesy as it is) a "fail" is a first attempt in learning—quite literally a fail! Using failure as stepping stones to the bigger picture allows us to grow and learn from our mistakes instead of giving up. Sometimes we are so tough on ourselves that we think we can't fail at what we try to do but in life, as in science, it takes a lot of failing and messing up to figure out the best way to tackle challenges. In our quest for perfection we often hear people say that failure is not an option, but in reality, experiencing failure is the only way to succeed in anything.

What is something you wish you did differently when you first started working?

As a current full time student, juggling research projects and extracurricular activities, what I realized quickly is how essential time management & getting proper sleep are so crucial to performance! I wish that I had started having a more organized schedule during the first semester instead of now, but I believe that everything worked out the way it was supposed to because that's how we learn—by experience!!! And these experiences make for great stories eventually!

If someone wanted to do what you do, what's the best piece of advice you'd share with them?

Just do it. Go for whatever it is that gives you a sense of purpose! I think it's important to advocate for yourself and tell yourself that no matter what, you can do whatever you set your heart and mind to. I think nothing is more empowering than believing in yourself first.

What scientific discovery or event in history would you like to go back in time and witness?

Maybe either Tycho Brahe's observation of supernova or the Wright Brothers' first successful flight or the Apollo Moon landing, I think that was such a turning point in human history where we realized that anything in and out of this world—is possible.

If you could definitively answer one unanswered science question, what would it be and why? / What unanswered scientific question keeps you up at night?

So many questions but this one keeps me up at night: How can we bend time and space and use that knowledge to enable interstellar travel for humans? Either that or questions related to cosmic inflation and black holes. I just had a class covering those concepts and it just blows my mind!

What's a misconception about your line of work?

That to be successful in the space industry, you have to follow a linear path and that's usually being either aerospace/mechanical/electrical engineering major. But as I move around and network with different industry professionals, I realized that there are so many career paths involved in this industry and what that tells me ultimately is that there are options you can take whether in science and technology or engineering, communications, business, law and even arts! So don't try to limit yourself and explore your capabilities and interests because there are many ways to contribute to this incredible field beyond being an engineer and scientist.

What's the best part of doing what you do?

Being in the forefront of research and development of both knowledge and technology as we push humanity forward towards an evolved future.

What is a necessary evil in your industry?

Competition. But I think it's a necessary evil because competition is also needed to be able to sift through the best ideas and solutions that could solve the hardest challenges we face. Healthy competition can bring out the best in people.

Have you ever changed your line of work? If so, why and what was the change?

Right now I'm still on track to pursue a path in the STEM/Space Science field. A little off road from my first aspiration to be a paleontologist but still in STEM!

What purchase under $100 has improved your life: career or personal?

Well it is valued at $250 but I ended up registering for free ($0) to attend a Higher Orbits' Go-For-Launch program back in 2019. And I think that's one of the best investment time or money wise that I've ever made as it changed my life, personally and professionally. Through that program, I was able to meet and work with my first astronaut mentor, Dr. Don Thomas, and Space mentor Michelle Lucas, as well as became part of a team that designed a winning science experiment, a payload actually, that would create art in space using microgravity. That invaluable experience launched my dreams and aspirations to forge a path towards the skies and beyond. So if you are in 8th-12th grades, love STEM and space exploration, and want the best bang for your buck, do check out: www.higherorbits.org and dive right into their non-profit educational programs like Go-For-Launch. The stellar opportunities that you get to have with Higher Orbits are definitely out of this world!

How has social media benefited or hindered your career?

I think social media, when used conscientiously, can be a positive force because it's a great networking place for various kinds of people, especially those that are in the same fields of interest as you. It can also be a great platform to advocate for causes and projects that you need to amplify on a global scale. I'm thankful for social media as it is the platform where I got to work with and meet so many incredible people in the space community like Kevin!

STEM PRO
JOAN MELENDEZ MISNER

JOAN MISNER: Mission Integration Systems Engineer, NASA Launch Services Program, MS Systems Engineering, BS Chemical Engineering & Chemistry

IG: *@YourFemaleEngineer*
Threads: *@YourFemaleEngineer*
FB: *Your Female Engineer*
YT: *@YourFemaleEng*
X: *@YourFemaleEng*
TT: *@YourFemaleEngineer*

JOAN MELENDEZ MISNER is a native of Orlando, FL., is a Mission Integration Systems Engineer at NASA for the Launch Services Program working on management of space and aeronautical flight systems for all non-crewed and scientific missions. Her missions include the Double Asteroid Redirection Test (DART), Europa Clipper, GOES-U and Dragonfly.

Joan received a dual bachelor's degree in Chemical Engineering and Chemistry from the University of Maryland, and a master's degree in Systems Engineering from Naval Postgraduate School. Prior to working for NASA, Joan worked for Blue Origin on the New Glenn rocket, and Naval Air Systems Command on jet engines, fuel systems, and biofuels research/ qualification.

In her spare time, Joan volunteers throughout the community and is fully involved with STEM Outreach Programs, including judging robotics competitions, mentoring middle/high school aged students, and participating in the "Women in Math" events. She was chosen as a "Wonder Woman" for the STEM-ing event due to her endeavor in the community. She was also chosen to be a part of the DoD's "30 under 30" promotional video which aimed at raising awareness about STEM career opportunities among college students studying in STEM.

As a first-generation college graduate, I strive to increase representation in underrepresented communities, as well as encourage them to pursue STEM careers through social media and nonprofits.

Additionally, she creates STEM content on both Instagram and TikTok under the name "YourFemaleEngineer".

STEM PROs Questions

Why do you do what you do? Do you have a defining/ah-hah/ eureka moment where you knew what you wanted to do?

Great question. I love being an engineer in the Space Industry because I feel like I not only get to work in an area I have dreamt of since I was a kid looking at the stars, but I also feel like I am making a small difference with the science missions I get to help launch.

I have a few (haha). One specifically that I remember was moving from Puerto Rico to Orlando. At that point, I was already fascinated with space and space travel. So, when my family moved to Orlando, I specifically remember one summer day our whole family was home, moving in, and our house shook violently. My mother thought it was an earthquake and quickly told us to go under the stairs for safety. After about 15 minutes or so, we came out and I turned the television on to see what caused the earthquake. Then, the news anchor said "Today we welcomed the 6 astronauts coming home from the Space Station. I'm sure we all felt that sonic boom". At that moment, I knew I had to work in the Space Industry!

What is something you wish you knew while in school?

I wish I knew that failure is a part of life. I am very open about my failures/ missteps on social media not because I love sharing it with the world, but I believe it's important we share the good and bad side about trying to achieve goals. This is especially important in the minority communities, most specifically first-generation students. As a first-generation college student, I had so much pressure already on me with being the first one in my family to go to college. So, when I failed my first exam in my STEM career, I took it hard, I thought that I was not good enough and the Imposter Syndrome started to creep in. I almost switched majors because I didn't want to fail my family and not graduate college. If I could go back to that young girl, I would tell her that everything will be alright. This is why I use my social media so much. I want others who feel lost or discouraged to know that we all felt that way at some point. We all have different stories/backgrounds. I had to work full time while in college, even though I had scholarships, because I had to help my family while I studied. I can't compare my journey with someone who may have been lucky enough to have going to college their full-time job.

My favorite quote from Arianna Huffington will always be "Failure is not the opposite of success; it is part of every success story".

What is something you wish you did differently when you first started working?

I wish I was more vocal. As a young engineer, I felt like I didn't have enough experience to speak up during meetings and make suggestions. It was a combination of being scared that my suggestions would be laughed at and not having the confidence in myself to speak up. That's why as a more seasoned professional, I always tell my young engineers or interns to never be afraid to speak up. I always make it a point to hear their suggestions and even incorporate them into projects. It gains them the confidence I never had when I started off.

If someone wanted to do what you do, what's the best piece of advice you'd share with them?

Network to Get Work. I am filled with so many quotes, but I resonate a lot with this one. When I first applied to NASA, I was rejected 13 times (I am sure Kevin knows this feeling too). I started contacting folks who worked at Kennedy Space Center through LinkedIn to ask questions. I also met a few at networking events and they ended up giving me excellent advice and even looked at my resume when I applied again. So, if you want my job or any other job, I suggest to talk to people in the industry whether through social media, LinkedIn, or networking events. Ask them how the job is like and if they are willing to advise you on how to get there. The worst someone can say is no or that they do not have the time. But most of the time, they can direct you to others who may have more time to help.

If you could definitively answer one unanswered science question, what would it be and why? / What unanswered scientific question keeps you up at night?

As a space enthusiast, my biggest scientific or just theoretical question is "Are we alone". I'm fascinated as to what is out there, and there is just so much to explore!

What scientific discovery or event in history would you like to go back in time and witness?

The Big Bang, but technically I can't witness that since we weren't created. But I am sure it was so beautiful when it happened.

More realistic, I wish I was present when we discovered the Higgs Boson back in 2012. As a chemistry major and a physics minor... this was absolutely incredible!

What's a misconception about your line of work?

The biggest misconception is the notion that space is only going to benefit the wealthy and the comment, "why are we going to space when we need help here on Earth".

While I never brush off these comments because I truly believe that we should take care of our planet and everyone in it, I also like to give examples of how space missions are helping us here without them knowing. For example, we launched SMAP back in 2015, and that mission's goal is to be an Earth orbiting satellite that measures soil moisture, which lets you predict floods, and crop monitoring (wheat, rice, and corn). Another good example is Sentinel-6b which was launched in 2020, which is collecting data on the depth of our oceans and getting us better data on how we can combat climate change. In both occasions, it is helping humanity in different ways and trying to preserve Earth. Most of the times, these missions are often overlooked, so when I explain the good we can do it helps with the misconceptions I mentioned above

What's the best part of doing what you do?

I love being able to work on something that feels as if it's bigger than myself. Working on science missions that go to space, I feel a sense of pride being able to collect data or explore other worlds/planets that can benefit humanity in the future. For example, working on the first planetary defense mission was so exciting, and being able to say I was a "Planetary Defender" is also kind of neat. Not only do we work on a mission like that, but collecting data that can help grow crops, combat climate change, or save a community from a weather event that can be destructive is something I pride myself on every day.

I also love the human side of what I do. What I mean by that is that I love being able to do outreach and inspire the next generation while "working". From going to schools or conferences, or mentoring and even bringing a student to work, I have so many different avenues to continue to show representation in the space industry through my job.

What is a necessary evil in your industry?

Hmmm, don't know how best to answer this question to be honest.

Have you ever changed your line of work? If so, why and what was the change?

I started my career in the aviation industry working on military fighter jets, qualification of biofuels, and aerial refueling. While I liked what I did, getting to work on cool aircraft, I knew that space was where I wanted to end up. In aviation I started working in the lab to then working more hands on with the jets, to then learning how to manipulate data and better improve maintenance intervals.

Now, in the space industry, my current role is a combination of technical and program management. So all of the roles I had in aviation has helped me succeed and become a better well-rounded engineer In my current role.

What purchase under $100 has improved your life: career or personal?

I love to read a lot and I can't pinpoint necessarily what specific book I have liked the most, so I will keep it general and say all books I have purchased. Whether it is to learn a new trade (finance, retirement, investing) or diving into a new series (Fourth Wing, etc.) I love to be able to escape reality and dive into a new book.

How has social media benefited or hindered your career?

It has done a little of both, but mostly benefited because I have had the opportunity to help so many start or navigate their career, and to me that alone is worth it.

In some cases, though, working for a federal company, it has sometimes hindered who I can work with and trying to navigate the perception of speaking for said federal agency. On my social media profiles I have to specifically state, "I do not speak for XX company and I am not affiliated with them". It can sometimes be frustrating because instead of hindering your employees and discouraging their use of social media, these companies should be embracing the fact that employees can reach an even bigger audience. Social media is a different means of outreach and the companies that can incorporate it into their business/agency will be very successful.

"However difficult life may seem, there is always something you can do and succeed at."

–Stephen Hawking, theoretical physicist, cosmologist, author, and recipient of the Presidential Medal of Freedom

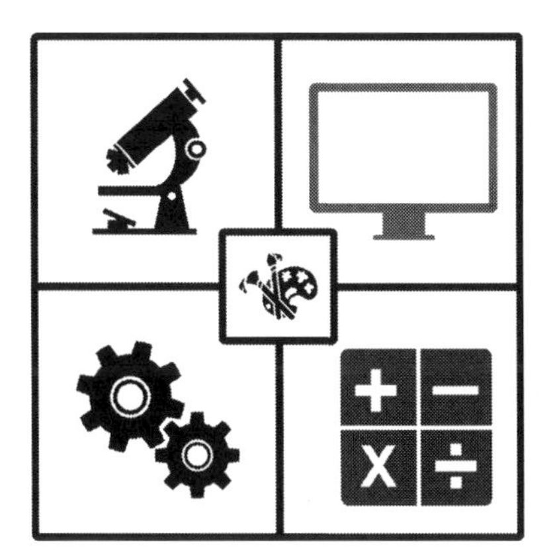

STEM PRO

JANELLE WELLONS

JANELLE WELLONS: Senior Mission Operations Engineer, ispace, inc., formerly NASA JPL

IG: *@itsjanellie*
FB: *Janelle Wellons*
LI: *@jwellons*
Website: *janellewellons.com*

JANELLE WELLONS graduated from the Massachusetts Institute of Technology with a B.S. in Aerospace Engineering, and she had no idea that it would spark the beginning of a 7 year career operating science instruments and spacecraft at the Moon, Saturn, and our own planet Earth. She is a Senior Mission Operations Engineer at ispace, inc. (as of March 2024), leading routine and critical space operations from the HAKUTO-R Mission Control Center as a Flight Director for the companies' lunar lander missions.

Previously, Janelle worked at the NASA Jet Propulsion Laboratory on the Earth observing MAIA, Sentinel-6, and SWOT missions as well as the Cassini mission to Saturn and Lunar Reconnaissance Orbiter. Speaking to students about pursuing careers in space has been a constant throughout her career and she has been grateful to share her message through features by PBS SciGirls, Nike, Lego, and more. At JPL she was awarded the Bruce Murray Award for "inspiring students to engage in STEM, quenching their thirst for knowledge, and sparking a curiosity greater than the stars in the sky."

When she isn't on-call to operate robots in space, you can find her exploring her new home in Japan and doing outreach for communities traditionally underrepresented in STEM.

STEM PROs Questions

Why do you do what you do? / Do you have a defining/ah-hah/ eureka moment where you knew what you wanted to do?

I never imagined that one day I would be operating spacecraft across our galaxy. Maybe a lawyer, a marine biologist, or a writer, but a space engineer? It seemed as far-fetched as wanting to be a princess when I grew up. By the time I graduated from high school and headed to my dream school, the Massachusetts Institute of Technology, my indecision on a career had not changed. I spoke to upperclassman for advice and they urged me to try one of the introductory courses on engineering that our school was well known for. The choices felt endless. I could try biological, electrical, mechanical, chemical, civil, even ocean engineering. But then there was Aerospace Engineering. Space. It was in the name. And while I couldn't tell you the first thing about rockets or planes outside of what I could see in the sky and stars above, I had always thought space and the study of our universe was awe inspiring. With that, I went to the first class and listened as the professor went over the syllabus. We would be learning about lighter than air vehicles, the rocket equation, what makes planes fly, and the history of spaceflight. He shows on the screen an image of an astronaut fixing the Hubble Telescope, free floating in the dark expanse of space. And then he casually remarks that he is the astronaut in the photo. This was my moment. My ah-ha. How many chances in a lifetime do you get to meet someone who has left our planet? And learn directly from them how it is done? It's all the convincing that I needed to put me on the path that has led to my incredible career in space.

What is something you wish you knew while in school?

That a career in space is not just about working on the flashy missions at Mars, Jupiter, Saturn and beyond. It's also about the great body of work being done to study our own home—Earth. This is the only planet we have and the ability to observe it from space is key to understanding the ways it is changing. If I knew this in school, I would have explored more of the classes centered around Earth science, remote sensing, and climate studies.

If someone wanted to do what you do, what's the best piece of advice you'd share with them?

Hone your leadership skills and emotional acuity to the same degree of your technical expertise. Flight Directors are the ultimate authority of the mission control center. When things go ary, the team turns to you for guidance. The ability to be the calm source of logic and direction in the midst of red alarms and panic is crucial to mission success. This is also the hardest thing to prepare for. You can train, rehearse, memorize, and study your checklists and procedures endlessly, but there is no stand-in to the feeling of watching your spacecraft separate from the rocket or perform its first orbital maneuver. The best thing you can do to plan for these moments is practice leadership through challenges wherever and whenever possible.

What scientific discovery or event in history would you like to go back in time and witness?

I would like to go back in time to see the first Moon landing. The context of this moment in history is just as important as that first step on lunar soil. It was the middle of the civil rights movement. Black Americans were fighting for their right to be seen as equals in a country that was built off their slave labor. The promise of taking humanity further than they have ever gone before contrasted starkly with an inability to show humanity to the country's own citizens. Protests abound while heavily funded teams led by NASA worked around the clock to make President John F. Kennedy's declaration to put man on the Moon a reality. And all the while, hidden from view were women working as human computers. Calculating the trajectory, creating the code, and practicing the engineering that would make this landing a success. It was a show of the best of us, accomplishing a feat that to this day has not been repeated off less computing power than a modern day cell phone. But it was also the worst, as racism ravaged a country that promised freedom and equality for some, not all. I think that to witness this would be to learn my own history and better appreciate the sacrifices of the men and women who came before and paved the path I currently walk.

What's a misconception about your line of work?

Many people assume that because I work in the space industry that I'm a rocket scientist. It's a title that is common in informal conversations but rare in my line of work. I don't work on rockets, I work on robots used for space exploration. And I'm not a scientist, rather I'm an engineer who operates those same robots on their missions. And while there are scientists who research and develop new ways to improve rocketry, they make up a small part of the potential career paths in space. The more you know!

What's the best part of doing what you do?

Being part of history. The Cassini mission to Saturn was my welcome to NASA JPL. Some of the people on the team had been working on the mission for longer than I had been alive, but they still treated me as one of their own. I was given the opportunity to operate the cameras that would capture breathtaking images of the planet, its rings and its moons. And when the Cassini spacecraft had run out of fuel and the Grand Finale began, I quickly learned that it was not just the end of the mission that we were gearing up for, but the end of a team and its legacy. The night we watched with bittersweet longing as its signal faded to nothing as it burned up in Saturn's atmosphere, pointing to Earth one last time, is something that I will always cherish in my memories. And while Cassini will be remembered for its historic contributions to space exploration, what I will remember it for is the team it brought together.

What purchase under $100 has improved your life: career or personal?

A good pillow is worth the money after a long night shift operating spacecraft.

How has social media benefited or hindered your career?

I mainly use social media to share my journey with people who may be wondering what the life of a space engineer may be like. I don't limit my posts to space content only. Instead, I share the highlights of my experiences living abroad, experiencing new culture, finding community, and giving back through outreach. In this sense I have felt that social media is not an obligation or burden but rather a way to document the important parts of my voyage for prosperity. It has also allowed me to connect with other space engineers from around the world who are pushing boundaries and making names for themselves.

STEM PRO

DR. TANYA HARRISON

DR. TANYA HARRISON: PhD in Geology with a specialization in Planetary Science and Exploration

IG: *@tanyaofmars*
TT: *@tanyaofmars*
X: *@tanyaofmars*
FB: *Tanya of Mars*
YT: *@tanyaofmars*
Medium: *@tanyaofmars*
Website: *tanyaharrison.com*

DR. TANYA HARRISON calls herself a Professional Martian. Over the past 15 years, she's worked in science and mission operations for multiple NASA Mars missions, including the Opportunity, Curiosity, and Perseverance rovers and the Mars Reconnaissance Orbiter. She holds a PhD in Geology with a specialization in Planetary Science and Exploration from the University of Western Ontario. She is the Co-Founder and CEO of the Earth and Planetary Institute of Canada (EPIC) and a Fellow of the University of British Columbia's Outer Space Institute. You can find her prolifically posting about all things Mars and her experiences as a scientist dealing with ankylosing spondylitis as @tanyaofmars on most social media platforms.

STEM PROs Questions

Why do you do what you do? / Do you have a defining/ah-hah/ eureka moment where you knew what you wanted to do?

I've been obsessed with space since I was about 5 years old thanks to things like Star Trek and the book The Magic School Bus: Lost in the Solar System, but it wasn't until NASA's Pathfinder mission landed on Mars when I was 11 that my interest honed in on Mars. The Pathfinder lander carried the microwave-sized Sojourner rover along with it, and snapped pictures of the rover driving off the lander platform and onto the surface of Mars. NASA released an animated GIF of these images online, back before animated GIFs were all over the place. When I saw this, I *knew* I had to work on Mars rovers. And I was laser focused on that interest from that day forward.

What is something you wish you knew while in school?

When I got to college, I went into astronomy because I thought, "Planets are in space, so that means I should be an astronomer if I want to study Mars!" But it wasn't until near the end of my junior year that I realized I actually should've studied geology. I don't regret my astronomy degree and the things that I learned in that process—it was supremely cool stuff—but I do wish I'd had the chance to take more geology courses before starting graduate school in geology.

What is something you wish you did differently when you first started working?

Honestly, I'm not sure there's something I would've done differently, because I managed to land my dream job straight out of my Masters program, so my approach to chasing that job paid off! As I was nearing graduation, I started cold emailing professors that had written the papers I'd read for my thesis research, to see if any of them might need a space data processing person in their lab. One of those professors recommended checking out a particular company that builds and operates cameras for NASA Mars missions. I never would've thought to look there on my own, as I would've assumed a company that builds cameras would be looking for engineers, not geologists. However, it turns out they wanted Masters-level geologists with familiarity of Mars to *operate* the cameras! And they just so happened to be hiring for an Assistant Staff Scientist to work in operations for cameras aboard NASA's Mars Reconnaissance Orbiter. I applied immediately, and probably borderlined on annoying in my enthusiasm in following up repeatedly to let them know I could start ASAP—all my stuff was packed and in my car, just say the word and I'd hit the road. The hiring process took a few weeks, but eventually I got the job! I ended up working there for about four years before deciding to go back to school for a PhD.

If someone wanted to do what you do, what's the best piece of advice you'd share with them?

Be proactive. Network with people both in person and online, share your genuine enthusiasm and your career aspirations, so that when opportunities arise, people think of you. And when those opportunities present themselves to you, be ready to jump on them. It's rarely as simple as scouring the internet for job postings and submitting a resume.

If you could definitively answer one unanswered science question, what would it be and why? / What unanswered scientific question keeps you up at night?

I want to know whether or not we're alone in the universe. Heck, are we alone in our own solar system? My dream is to live long enough to see a mission make it through the icy shell of an ocean world like Europa or Enceladus, turn on its lights, and *bam*—we see water full of cosmic krill and space whales. Something bigger than a microbe. Something unmistakably *alive.*

What scientific discovery or event in history would you like to go back in time and witness?

I wish I could see the early days of robotic planetary exploration, like the day Mariner 4 flew past Mars for the first time, or Voyager 1 flew past Jupiter. We take stunning photos of the planets and our universe at large for granted these days, being inundated with a flood of data from telescopes and rovers and satellites whizzing through space. But it wasn't all that long ago that we'd never seen another planet close-up. Mariner 4 sent back the first close-up images of Mars in 1964—within my dad's lifetime! Before that, there were still dreams of alien civilizations on the surface. It's wild to think about just how much we've learned in the span of a single generation. We probably can't even fully dream of what's to come in our generation.

What's a misconception about your line of work?

The biggest misconception is probably that we're sitting in Mission Control driving rovers on Mars with joysticks like it's Mario Kart. Not to burst anyone's bubble, but it's visually much less exciting than that. Since there's a time delay of anywhere between 4 and 40 minutes to send signals to/from Mars, it's not practical to operate the rovers in real time. So, we write a bunch of code to tell the rovers what to do for an entire day, send that off all at once, and then sit back with our fingers crossed hoping the code and the rover all behave properly! Usually it's fine, but sometimes Mars throws a curveball, like a dust storm or some sand that doesn't play nice with the wheels. That's when you have to start getting creative.

What's the best part of doing what you do?

When it comes to space missions, you never know what you're going to see day to day. You're constantly learning, constantly exploring. It's an exhilarating experience, even on the mundane days. Even after years of working on a single mission, I'd still practically dance my way into work with excitement about what I might see in the images the cameras sent back overnight. Oftentimes, I was the first human on the entire planet to see those images. That's an indescribable feeling. For a brief moment, you have a beautifully intimate relationship with another planet.

What is a necessary evil in your industry?

There are a lot of ties between the space sector and the military industrial complex. A lot of space companies in the U.S. for example get huge amounts of funding from the Department of Defense. That means that defense applications overwhelmingly drive the direction of innovation—not science. As a scientist, this harsh reality can be depressing. It makes me wonder how much more we could do, how much more we could know about our universe, if scientific exploration drove more of the innovation when it comes to the space sector. We're also seeing a big push into space companies focused on resource extraction from places like the Moon and asteroids, which feels a bit worrisome given the history around resource extraction here on Earth. It's sold to us as the answer to reducing extraction on Earth, by moving mining offworld. But, if we require resources beyond our own planet to support our civilization, is that not the antithesis of sustainability?

Have you ever changed your line of work? If so, why and what was the change?

In early 2023, I left what had been a dream space job due to bullying and harassment at a well-known space company—and even more so their handling of the situation than the harassment itself. Instead of jumping immediately into a new job at another space company, I decided to take some time off and re-evaluate how I wanted to integrate space into my life in a way that brought me joy again. After a lot of time and reflection, I decided that the best way to return to space would be to create the work environment that I wanted to see for myself and others in the industry. I needed to start my own thing, rather than going back into another corporate setting that could turn toxic.

What purchase under $100 has improved your life: career or personal?

A physical notebook small enough to carry with me at all times, combined with a fountain pen. It both gives me the opportunity to write down my thoughts anytime, anywhere, while also forcing me to do it with intention because the pen takes a nonzero amount of time to dry as you write. In my sabbatical from space in 2023, I wanted to disconnect from the digital world and slow down, and this pen and notebook combo was one fantastic way to do that. I've already crushed through an entire notebook and just recently purchased a second to continue the physical writing process.

How has social media benefited or hindered your career?

Twitter absolutely launched my career in a direction that I never could've expected. I never set out to be a social media personality or influencer intentionally; the following just kind of happened over time as I started moving into sharing more of my personal story and journey rather than just talking about Mars. It led to job offers, speaking gigs, TV appearances, and more, as well as gave the chance to connect with tons of other scientists and science enthusiasts around the world. It breaks my heart that Twitter as it used to be no longer exists, and that the same vibe and momentum hasn't really been replicated elsewhere (yet?). Science Twitter was a special place, and I'm very grateful for the opportunities it gave me.

STEM PRO

MARY NGUYEN

MARY NGUYEN: Software Project Manager | Ex Spotify, MDA, & NASA

IG: *@marynguyenco*

MARY NGUYEN is a former engineer at NASA and Missile Defense Agency (MDA), where she then transitioned into Tech as a Project Manager. Throughout her career, she's showcased the successes of being a multifaceted woman while thriving in a male-dominated field. She was the founder of @shopmnco, a clothing line that started from her love for both fashion and STEM. She is a first-generation college graduate and second-generation immigrant, which has been represented throughout her brand. Her platform shows snippets of her lifestyle while you can also find professional insights and resources through her past public interviews and blogs.

STEM PROs Questions

Why do you do what you do? / Do you have a defining/ah-hah/ eureka moment where you knew what you wanted to do?

When I was an engineer, I loved problem solving and bringing concepts to life. It was technical yet creative, because I was able to conceptualize and deliver products based on technical requirements. My "ah-hah" moment was realizing that I enjoyed a good challenge and figuring out how things worked and why. Coupled with that was my love for working with people and being in a more client-facing role, which is why I became a Project Manager. I don't do the technical work anymore, but it helps having that background. I still problem-solve and come up with creative solutions, just in different ways!

What is something you wish you knew while in school?

It's something that I eventually figured out—the classroom doesn't always equate to the real world. You have to get out there and do it.

What is something you wish you did differently when you first started working?

Ask more questions and be more proactive with my teammates. I can be too self-reliant and that's not always a good thing.

If someone wanted to do what you do, what's the best piece of advice you'd share with them?

Determine your end goal and work backwards. Look up similar professions and see what the job requirements are, then determine how to obtain those skills. Your hands-on experience and network will get you farther than your GPA. Make yourself marketable and competitive.

If you could definitively answer one unanswered science question, what would it be and why? / What unanswered scientific question keeps you up at night?

How did life begin? What happens when you die? Are we alone in the universe?

What scientific discovery or event in history would you like to go back in time and witness?

Why can't I think of any?!

What's a misconception about your line of work?

When I was an engineer, it was that you had to be crazy smart or in general, that there's only one way to get to where you want to be. The one thing I love about STEM is that your skills can transfer. As we professionally evolve, our interests may also evolve, and we can make career shifts. I love that we can reinvent ourselves as many times as we want, there are no rules.

What's the best part of doing what you do?

It's rewarding and impactful. I help deliver innovative products and it's awesome being able to support the success of others, whether it's the client or business. My job is also remote first, as many Tech/Software jobs, and the flexibility is unbeatable.

What is a necessary evil in your industry?

Administrative tasks, yuck.

Have you ever changed your line of work? If so, why and what was the change?

I graduated with a B.S. in Mechanical Engineering, kicked off my career as an engineer for several years, and then worked my way into Project Management.

What purchase under $100 has improved your life: career or personal?

A journal and second monitor, I use both consistently for career and personal.

How has social media benefited or hindered your career?

Social media has definitely benefited my career and connected me with so many amazing opportunities and people that I'm thankful to call my friends. I hope those needing a sense of community are able to find that by sharing pieces of their journey and stories.

STEM PRO

DR. INDIRA TURNEY

DR. INDIRA C. TURNEY: Black Caribbean Neuroscientist; Earl Stadtman Tenure-Track Investigator at the National Institutes of Health (NIH)

X: *@indiraturney*
IG: *@indiraturney*

DR. INDIRA C. TURNEY, a Black Caribbean Neuroscientist, earned her bachelor's in psychology from the University of the Virgin Islands and her Ph.D. in Cognitive Neuroscience from Pennsylvania State University. She is currently an Earl Stadtman Tenure-Track Investigator at the National Institutes of Health (NIH), directing the Brain Health Equity (BHE) Unit within the Laboratory of Epidemiology and Population Sciences (LEPS) at the National Institute on Aging (NIA). Her research focuses on understanding the multidimensional factors influencing brain health in diverse Black adults. By evaluating environmental, sociocultural, and biological pathways, Dr. Turney aims to achieve equitable brain health outcomes for all individuals, with a particular emphasis on historically marginalized communities. Dr. Turney is not only dedicated to advancing research but also passionate about fostering inclusivity in the biomedical sciences. As a co-founder of the Women of Color Writing Accountability

Group (WOCWAG) w/ Black Women PhDs®, she's fostered a community where WOC, across industries, support each other through writing accountability, meditation, social events, speaker series, and much more.

STEM PROs Questions

Why do you do what you do? / Do you have a defining/ah-hah/ eureka moment where you knew what you wanted to do?

I still don't know for certain what I want to do. What I do know is that I follow my passion for being in service to my community, whether through research, mentoring, or other paths. As long as I feel like I am giving back, I know I am doing what I am called to do. An unforeseen journey of exploration has led me here, allowing me to express my passion for understanding brain health equity across diverse populations, inspired by my desire to serve my community through science.

What is something you wish you knew while in school?

I wish I knew that my choices to become whoever I wanted to be were limitless. Having a mindset of limited options made me think I needed to be X to fit into X, rather than exploring career paths that fit my passions and skills. Thankfully, I caught on eventually!

What is something you wish you did differently when you first started working?

I wish I had taken significant time to transition between career stages. I find it so helpful to intentionally close one chapter before I begin another.

If someone wanted to do what you do, what's the best piece of advice you'd share with them?

Explore the options that call to you. You don't have to figure it all out now. Pursue your passion, seek mentors, and share your dreams and aspirations with those who will listen and encourage you. Surround yourself with people who inspire you (reach out to me)! Neuroscience is very interdisciplinary; you can do almost anything with a PhD in Neuroscience and related fields, so find your calling within the field, and know that it may not already exist—that's why we need you.

If you could definitively answer one unanswered science question, what would it be and why? What unanswered scientific question keeps you up at night?

Everything is connected, but how? Does it matter?

What scientific discovery or event in history would you like to go back in time and witness?

I would like to witness the creation and documentation of the Edwin Smith surgical papyrus around 1700 BC. This ancient Egyptian text provides the earliest known reference to the brain and detailed descriptions of various medical conditions and treatments, marking a significant milestone in the history of medical science. Observing this process firsthand, learning from the individuals who carried out these early medical practices, and understanding their hypotheses and methods would be fascinating. It would reveal how their unique approaches, though not considered scientific by today's standards, laid the groundwork for modern science. This experience could also encourage us to be more open-minded in our data collection and study of human health. Overall, this event would offer invaluable insights into early medical practices and the origins of neurology.

What's a misconception about your line of work?

A common misconception about being a neuroscientist is that our work solely revolves around brain/MRI scans. While brain imaging techniques like MRI and EEG are indeed valuable tools in our research arsenal, they represent just one facet of the diverse methodologies employed in neuroscience. Neuroscientists utilize a wide range of techniques and approaches to study the complexities of the brain. These include, but are not limited to, electrophysiology to measure electrical activity in neurons, molecular biology techniques to investigate genetic and biochemical aspects of brain function, and behavioral experiments to understand how the brain controls behavior. Additionally, we explore the impact of life experiences on the brain, analyzing how different environments and events shape neural pathways and influence brain function. Computational modeling is also used to simulate brain processes. By employing this multifaceted approach, neuroscientists can gain a more comprehensive understanding of the brain and its functions, considering both biological and experiential factors.

What's the best part of doing what you do?

The best part of my work is the wide range of things I can do as a career as well as my day-to-day. I've worked as a scientist in an academic/ government setting, others pursue careers in industry, government, healthcare, or nonprofit organizations. My day-to-day tasks may include, writing up scientific findings for publications, presenting findings at scientific conferences, as well as lecturing at universities or other settings, travel around the world disseminating my research and collaborating with other scientists. I also thoroughly enjoy the opportunity to mentor peers and trainees at all levels.

What is a necessary evil in your industry?

In neuroscience, a necessary evil could be the complexity and unpredictability of brain function, which often complicates research and treatment outcomes. This unpredictability makes it challenging to develop effective treatments and interventions. It also underscores the importance of meticulously understanding the underlying mechanisms and causes of changes in brain structure and function, as well as considering the impact of individual life experiences on brain health. Balancing these complexities is crucial for advancing our knowledge and improving patient care.

Have you ever changed your line of work? If so, why, and what was the change?

Partially. I transitioned from general cognitive neuroscience research focusing on memory in aging to now focusing on brain health equity among diverse populations. This shift was driven by a desire to address disparities in brain health and ensure that research benefits a broader segment of the population.

What purchase under $100 has improved your life: career or personal?

A subscription to a meditation app for enhancing mental well-being, benefiting both my personal and professional life.

How has social media benefited or hindered your career?

Social media has benefited my career by allowing me to make meaningful connections, share my academic journey, and foster mentorship relationships. It has also provided access to job postings and professional development opportunities. However, it can hinder productivity through procrastination and the spread of misinformation, leading to misunderstandings or misconceptions about scientific research. Combatting misrepresentation of work on these platforms is also a significant challenge.

STEM PRO
MAKIAH EUSTICE

MAKIAH EUSTICE: Aerospace Engineer, US Air Force Officer, BS Aerospace Engineering, Analog Astronaut, Streamer

IG: *@astro_eustice*
X: *@astro_eustice*
YT: *@astro_eustice8881*
LI: *@makiaheustice*
Twitch: *Astro_Eustice*

MAKIAH EUSTICE is an engineer and officer in the US Air Force. She has an intense passion for human spaceflight, and her astronaut dreams have led her to carve unique paths for herself and apply the education to bettering the human experience on Earth.

At the time of this writing, Makiah worked as a flight test engineer for 3 years. She has planned and executed tests for modifications on cargo aircraft, like the C-130 Hercules and C-5 Super Galaxy. Her undergraduate degree was in Aerospace Engineering where she worked on shape-memory alloy applied research.

As an analog astronaut, she has lived, worked, and researched in 3 simulated space missions for a total of 27 days. Her latest one in January 2020 was at HI-SEAS, Big Island, Hawaii, where her 5 female crewmates explored lava tubes on Mauna Loa, ate freeze dried food, and managed limited water and electricity for 2 weeks.

In her undergraduate time, Makiah created focus groups in human spaceflight and cubesat design while growing the Students for the Exploration and Development of Space (SEDS) Chapter to be inclusive to all majors. She was adamant to do the extraordinary beyond the curriculum of her degree.

Her tenacious attitude led her to be selected as a Brooke Owens Fellow, attend International Space University, train for mesospheric suborbital research, and attend NASA Social events for east and west coast launches. She is also a variety aerospace streamer, playing aerospace-themed games, crafting, and talking about space news. In her free time, Makiah loves to lift, watch cartoons with her cat, play music, and garden.

STEM PROs Questions

Why do you do what you do? / Do you have a defining/ah-hah/ eureka moment where you knew what you wanted to do?

I feel like I stumbled into a passion for STEM, and later space. One day I was reading the highschool course descriptions and was fascinated by the subjects taught in Advanced Placement (AP) Physics. As I excelled in math and science, I joined the robotics team and had my first experience with power tools, coding, and controllers. I started to hear about "new space", like SpaceX and Virgin Galactic, setting huge visions for humanity beyond Earth. It shook a fundamental desire in my soul to contribute to something humanity changing. Of course, I set out to do Aerospace Engineering in order to be a rocket scientist! Towards the end of my college freshman year, another path revealed itself. I was in the Air Force R.O.T.C. program while being in the Texas A&M Corp of Cadets program. I learned that the Air Force wasn't just pilots, there were engineers too. My curiosity carried me all the way to a jet flight, riding in the backseat of the T-38 Talon trainer. That summer, I went to Adult Space Camp, meeting long-time fans of the Shuttle, carrying out spacewalks to repair a malfunctioned satellite, and directing mission control. I came back knowing I wanted to not just be the rocket scientist, but the rocket rider!

What is something you wish you knew while in school?

"I wish I knew how to study" may be a common answer and one I connect with. Really, I wish I knew how my brain worked! Struggling to focus or break down a problem, I'd think I was just lazy or didn't have the aptitude for engineering. I wish I had given myself more permission to fail successfully. Ask "dumb" questions, don't stay mentally stuck in my head. I shouldn't expect myself to get things on the first few tries, it takes practice and exposure (over and over again) to understand and utilize concepts. I worked better with classmates that I felt safe to be "dumb" around, and we'd talk it out. Your self-efficacy is so important; you need to believe in yourself more than anyone.

What is something you wish you did differently when you first started working?

It was very difficult, like for many, beginning my career near the start of the pandemic. I was eager to be a sponge and be firehosed with lessons and technical exposure as a new Lieutenant. Instead, I was sent home with my laptop after a month. While I did learn, I was often bored, stuck, or left feeling aimless without any structure in my day besides Teams calls (the only chance to see other peoples' faces). With the lack of friends in a new state, I developed close online friendships and personal support. I would have benefited from building a professional or academic support network during that time. Despite the large amount of free time, I did not feel mentally ready to start a Master's program, but I wish I would have sought after certifications or learning groups to build up my skills. It took me a while (years) to realize that I work best with external factors, like working among others or clear intermediate goals. I'd encourage anyone to remember that how you start is not how you end your journey;

If someone wanted to do what you do, what's the best piece of advice you'd share with them?

Some people say that your connections are going to get you farther than your degree. I believe that applies in the aerospace industry and even the military. If you are studying for your degree or working already, prioritize excelling in that and making a great impression, but still build your network outside. Ask around, let your professor or supervisor know your vision. I moved into flight test engineering early because my Colonel knew what I wanted to do and someone needed to swap out positions. I got a scholarship to attend the International Space University in partnership with the Brooke

Owens Fellowship. I didn't make stellar grades in undergraduate, but I wouldn't have gotten this far without advocating for myself and investing in the relationships in my community.

If you could definitively answer one unanswered science question, what would it be and why? / What unanswered scientific question keeps you up at night?

I'm always thinking about how we can have a more circular economy and properly reuse Earth's valuable materials. Plastic, despite its environmental impact, will always be around in our lives. I'm interested in bioremediation methods for plastic that will bring down degradation time from centuries to months. I'd like industrial bioremediation facilities to replace trash dumps. There is research on certain bacteria and fungi that have enzymes that can break down certain types of plastic. Every once in a while I read a paper at 3am, even though I'm way out of my depth of knowledge. I might have to relearn some biochemistry.

What scientific discovery or event in history would you like to go back in time and witness?

I'd love to be able to see the first humans to step on another celestial body, the moon landing on July 20th in 1969, on a little black and white tv. I do believe I will get to witness the next steps on the moon in a few years.

What's a misconception about your line of work?

As an engineer in acquisitions, there will not always be much "engineering" work. The technical degree gives you skills to absorb the technical aspects and use proper judgment. There may be jobs where you are writing lines of code, setting up a test apparatus, or analyzing signal data, but more often than not you will be depending on contractors and civilians to do so.

What's the best part of doing what you do?

Despite only coming to the Air Force to do Flight Test, I'm seeing that there are many other cool opportunities if you have the right skills and stick your neck out. You get the advantage of seeing many different programs and roles while also keeping job security and going to neat locations.

What is a necessary evil in your industry?

The slow pace of government work sometimes can't be avoided. As you come in, excited to see planes being built or satellites launched, there is a lot of bureaucracy and red tape to work through. It makes sense, in that what we do is beholden to the taxpayers, but it can often be frustrating.

Have you ever changed your line of work? If so, why and what was the change?

My first job was as a structural engineer in an aircraft program office. I switched after a year to do flight test engineering, which was my driving interest to join the Air Force. Now I'm in contract management. Whether I keep pursuing the flight test engineering route and pursue Test Pilot School depends on my academic, financial, and personal goals over the next few years. I still long to work AND participate in human spaceflight, and the opportunities seem greener on the other (civilian) side. I don't want 20 years to pass by with comfort but regret what I could have contributed to. I believe, if I allow myself to be open to growth and to let go of things that no longer serve me, fulfilling work will always follow.

What purchase under $100 has improved your life: career or personal?

The only purchase I can really think of that has made my everyday life better… is an attachable bidet. I seriously had my doubts for so long, however, I can't see regular American toilets the same anymore. I truly believe they belong in every household. Need I say more?!

How has social media benefited or hindered your career?

Social media has been my port of discovery and ambitiousness for many years. I was quick to start sharing my experiences on Instagram of college and my aerospace journey. Yes, I look back and laugh at my cheesy pictures and word choices, but my enthusiasm was unmatched. I could practice audacity I rarely could muster in person. Somehow I ended up messaging a former astronaut "I'm making a student group for aspiring human spaceflight professionals/astronauts, can I get some advice?" Flash forward and this person is writing my recommendation letter for competitive schools. I was able to see my first live rocket launch out of Vandenberg AFB, as part of the NASA Social program, where I was given media access to pre-launch and launch activities of the ICESat-2 mission. It's where I get introduced to many new ideas, perspectives, and areas of expertise that I can be a student of, even while scrolling. However, I have a healthy respect for the psychological drawbacks of social media. I find myself comparing myself to peers, people my senior, people my junior; I feel the echoes of the multiverse of all the things I "should " or "could " have done by now. That's when it's time to step away, look at the ground I laid, and mold clay. Let my audacity live offline too.

"If you take a look at the most fantastic schemes that are considered impossible… you realize that they can be possible if we advance technology a little bit."

–Michio Kaku, theoretical physicist, co-founder of string field theory, and author

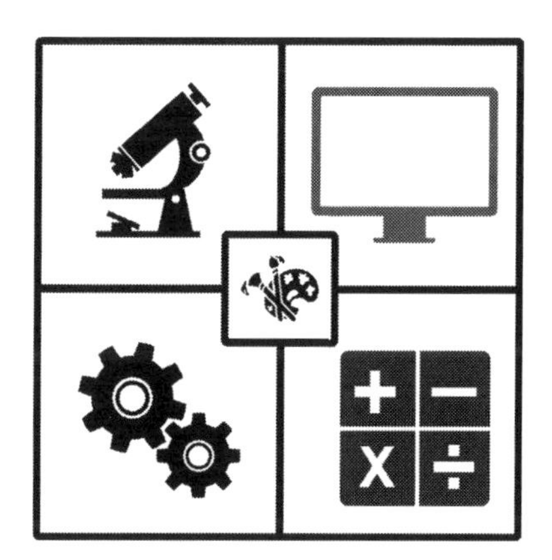

STEM PRO

JESSICA ASHLEY GAGEN

JESSICA ASHLEY GAGEN: BEng Aerospace Engineering, Fashion Model, STEM Advocate, Miss England

IG: *@jessicaashley_*

JESSICA GAGEN is an internationally signed fashion model and holds a bachelor's degree in aerospace engineering from the University of Liverpool. During the first year of her studies she noticed a lack of females in her cohort, and so made it her mission to advertise the subject and the many pathways STEM can lead to, to young women and children. Her efforts saw her scouted for the Miss England competition, which she won in 2022, and she has since used the platform to expand her mission with help of international media outlets.

Jess regularly creates STEM content and conducts her free #STEMSchoolTalks across the UK, which has led her to collaborate with Boeing, the ISS National Lab and speak at the Ascend Space Conference, as well as in many respected organisations—including the UK House of Lords and the Royal Aeronautical Society. In 2024 Jessica competed at the 71st Miss World contest, where she used the international spotlight to further her STEM mission. Jessica managed to win the coveted public speaking round after delivering a speech about decarbonising aviation through hydrogen, and left the contest with the title of Miss World Europe, having placed highest out of the contenders from the European continent.

Jess remains extremely passionate about showcasing STEM, and plans to do so through her links to UK media. In time, Jess would like to return to education to complete a Masters in Advanced Aeronautical Engineering, begin a professional career in engineering and use her links to television to present educational TV documentaries, set to inspire a new generation of scientists and engineers.

STEM PROs Questions

Why do you do what you do? / Do you have a defining/ah-hah/ eureka moment where you knew what you wanted to do?

Science and engineering lead the future of humanity, and realising how few women there were in my classes at university, I felt responsible to empower the next generation and show young girls that they don't have to lose their femininity to become fantastic engineers and scientists!

What is something you wish you knew while in school?

Whilst in school I wish I was more prepared for the workload and how fast the time passes.

If someone wanted to do what you do, what's the best piece of advice you'd share with them?

To create your own pathway. Sometimes in life you won't know of anybody doing that thing you dream of doing—be the first.

If you could definitively answer one unanswered science question, what would it be and why? / What unanswered scientific question keeps you up at night?

I am very interested in technology involving propulsion, so I'd be interested to learn what we could create which would let us travel at the speed of light.

What scientific discovery or event in history would you like to go back in time and witness?

The first moon landing in 1969!

Have you ever changed your line of work? If so, why and what was the change?

Yes, I started working as a fashion model from the age of 15 and had a successful career across the UK, Europe and Middle East. I decided to study Aerospace Engineering when I was in my early twenties because aside from being super interesting, I knew the degree would give me a solid foundation of skills employable in lots of different industries.

What purchase under $100 has improved your life: career or personal?

I love to read and run, so all of my books and my running shoes.

How has social media benefited or hindered your career?

Social media has given me a platform to connect with professionals in industry and further expand my STEM campaign, so I believe it has helped my career.

STEM PRO

CHEYENNE SMITH

CHEYENNE SMITH: Arts Worker, Aspiring Astronomer, @spaceforusofficial Founder

IG: *@blackspacegirl,*
@spaceforusofficial
TT: *@blackspacegirl*
FB: *Space For Us*
YT: *@spaceforus*
Website: *space-forus.org*

CHEYENNE SMITH is an arts worker, citizen / aspiring astronomer who founded Space For Us (@spaceforusofficial), an emerging nonprofit with a mission to bring visibility and accessibility to space sciences and industry for underrepresented communities. Their initiatives also encompass dark sky protection. Currently, Cheyenne is building a mobile observatory, set to be featured in local and regional neighborhoods, as part of Space For Us ongoing efforts.

STEM PROs Questions

Why do you do what you do? / Do you have a defining/ah-hah/ eureka moment where you knew what you wanted to do?

I always knew I wanted to be an astronomer since I was in elementary school. My ah-hah moment came when I saw the Hubble Ultra Deep Field image inside an encyclopedia given to me by my mom's colleague who was also a nerdy person. When I saw that image I wanted to know more about our place in the universe and find out if we are alone or not. The outreach and community work came later when I realized how it was missing in my local community and just how important it is for everyone to be represented.

What is something you wish you knew while in school?

That everyone has a math brain and to approach it like learning another language. Also, seek community!

What is something you wish you did differently when you first started working?

To start my own organization from the beginning even knowing how challenging it would be.

If someone wanted to do what you do, what's the best piece of advice you'd share with them?

Failure is always an option, giving up is not. It's okay to pivot! And it's never too late, you can always change your mind and pursue your dreams in life.

If you could definitively answer one unanswered science question, what would it be and why? / What unanswered scientific question keeps you up at night?

This is so hard!!!! I'll just go with what's been keeping me up lately, which is how we can describe the universe with one grand unified theory and test it.

What scientific discovery or event in history would you like to go back in time and witness?

This is hard too!!! I'm unsure. But I'd like to take our technology today and travel back in time to meet all the brainiacs back in the day and combine thoughts and ideas using the latest in tech today.

What's a misconception about your line of work?

That these spaces are meant for one demographic of people with a certain level of intelligence or "genius" brain. It only takes being curious, hard work, and the passion to learn more!

What's the best part of doing what you do?

I get to support others in whatever capacity I can. I get to provide an outlet that I wish I had growing up. When others in my community win I feel like we all are winning.

What is a necessary evil in your industry?

Math and pushing through academia / a PhD.

Have you ever changed your line of work? If so, why and what was the change?

YES! When I was discouraged from pursuing astrophysics in college, I went to major in something else I thought I was good at, which was communications (as an introvert this is a bit ironic). But I don't really like writing lol. I always loved researching growing up. What that looks like professionally is something I'm still in the process of learning. However, in my previous work I didn't feel fulfilled. Doing the work I do today fulfills me. It's a physical feeling I get.

What purchase under $100 has improved your life: career or personal?

Pencil and notebook. Writing things down, even if it doesn't make sense, helps me with some anxiety I feel about work and personal life stuff.

How has social media benefited or hindered your career?

Social media has helped me develop a community with experts in space sciences and industry. Something I rarely see in my hometown. This is primarily why I got on social media. It hinders me because I don't know how to tell my story in a compelling way. If I'm being honest, I don't really care to share much. It's not something that I necessarily want to do but have to because it's part of making this work more visible and relatable. I just never had a true affinity for social media and that part of me lingers / shows on my accounts. I'm not good at keeping people updated via posts and reels. I wish I was better though.

STEM PRO

SANDRA LAPLANTE

SANDRA LAPLANTE: Founder @papaytutor, Democratizing Access to High Quality Tutoring

IG: *@sandrav.laplante*
TT: *@papayatutors*
FB: *Papaya Tutor*
YT: *Papaya Tutor*

SANDRA LAPLANTE is a latina engineer who has tackled Silicon Valley's toughest challenges, an immigrant advocating to close the education gap, and an idealist working to level the playing field of STEM with her social driven startup, Papaya @papayatutor.

Sandra LaPlante, the CEO and Founder of Papaya, emerged from a personal struggle to afford STEM tutoring, a challenge that would shape her trajectory. Faced with financial constraints, she ingeniously turned to outsourced tutoring in Peru, a decision that not only bridged her educational gap but propelled her to graduate with an Engineering Degree. Post-graduation, Sandra entered the corporate arena, contributing her talents to Accenture and consulting for industry giants such as Google, Meta, Intel, and Starbucks. In these roles, she immersed herself in corporate strategy, technology, and software automation, gaining invaluable insights and expertise. Known for her collaborative approach, data-driven mindset, and unwavering customer focus, Sandra became a proven force in delivering high-impact solutions. Her consultancy extended to partnering with C-level executives, offering strategic problem-solving for Fortune 500 companies spanning diverse sectors, including Media, Technology Products, and Aerospace. The multifaceted experiences accumulated throughout her corporate journey converged in a pivotal moment, inspiring Sandra to establish Papaya. Papaya Tutors, the brainchild of Sandra LaPlante, is an online tutoring platform designed to democratize STEM education. Driven by a commitment to affordability and accessibility, Papaya empowers underserved students to strengthen their grasp of STEM subjects, unlocking their full academic potential. Sandra's entrepreneurial vision, coupled with her extensive technology and strategic consulting background, is the driving force behind Papaya's mission to make quality STEM education accessible to all.

STEM PROs Questions

Why do you do what you do? / Do you have a defining/ah-hah/ eureka moment where you knew what you wanted to do?

When I was working in Silicon valley I realized that there was no diversity in big companies because usually black and brown people come from low-income families and don't have access to quality education. So I realized the gap and I wanted to fill that gap and I said to myself, well, I was able to access high quality mentoring because I outsourced it. So, I'm going to build a platform that can help students access high quality tutoring for a fraction of the cost.

What is something you wish you knew while in school?

Actually I would have liked to know three things:

1. I wish I had known that when you graduate, you won't always work on what you study, many times you don't even talk about the subjects you studied during your college career. So, the more we learn about everything, the more opportunities we will have in the future.On the other hand, I wish I had asked more questions of my professors when I was a student.
2. The second would be discipline is the best form of self-love for me.

3. Ask for help, your teachers are there to help you, they are human just like you, if you fail, they will guide you, it is better to fail in school than in the real world.

What is something you wish you did differently when you first started working?

To be more confident, to understand that I don't have to have all the answers and that people should know, but to, to know what I know and know what I don't know and not pretend to know it all or understand it all and ask more questions.

If someone wanted to do what you do, what's the best piece of advice you'd share with them?

I would say that having a start up, it's very hard and don't do it just for the sake of having a start up but do it for love, do it because you want to solve a real problem.

Find a real problem and solve a real problem. So first you have to know your customer and then build your solution or build a parallel one because you have to build for people, not for what you think people need, but build for what people need.

If you could definitively answer one unanswered science question, what would it be and why? / What unanswered scientific question keeps you up at night?

How can we leverage emerging technologies and cognitive science to ensure personalized learning experiences that address the unique needs and learning styles of every student, regardless of socio-economic background or geographic location?

What scientific discovery or event in history would you like to go back in time and witness?

I think that I would have liked to see Albert Einstein. I would have loved to go to one of his lectures when, you know, he was explaining acceleration. I would love to hear from him because he had a very prophic mind, I think he pictured things.

What's a misconception about your line of work?

People think that you are going to do what you went to school for.And mainly it's not, it's just problem solving.I think people hire engineers to throw them into problems. And I think that that translates well into being an entrepreneur and a founder, right? It seems like 40% of us and you can see 40% of entrepreneurs in the world that are the richest people in this world happen to be STEM people.

What's the best part of doing what you do?

Helping more people to believe in themselves again and reach their potential, I think that is one of the most satisfying things to help, I wish someone would have done the same for me when I was a student. Many times school ceases becoming our safe place and we get scared, but at Papaya we work to make you feel good while learning.

What is a necessary evil in your industry?

That we are workaholics and it is not because it is mandatory, the point is that we feel so motivated to make a change that will help improve the confidence and learning of students who may not have imagined they could choose a complex career because that is how they perceived it at the beginning, but now they see complex subjects as easy.

Have you ever changed your line of work? If so, why and what was the change?

When I founded Papaya, I left the world of corporate consulting engineering practice behind because I felt a calling to help improve online education for underprivileged students.

What purchase under $100 has improved your life: career or personal?

I would say books! In my free time, I try to read something that makes me grow and opens my mind. I consider a book to be one of the best investments we can make in our lives as we are taking years of experience from someone down to hours.

How has social media benefited or hindered your career?

I think social media has helped me to expand and network with my associates and team. I usually use linkedIN more than other social networks and I am glad to see how more people want to join our team or see how we have a community of raving fans that constantly interact with our content.

STEM PRO

EMILIA ANGELILLO

EMILIA ANGELILLO: former microbiologist, multi awarded Science Lab Technician and STEM Ambassador, Sancton Wood School, Cambridge

IG: *@emilia.science*

EMILIA ANGELILLO is an RSCiTech Science Technician and a STEM Ambassador at Sancton Wood School in Cambridge. She completed her studies in Italy, earning a Bachelor's degree in Biology and a Master's Degree in Microbiology and Virology. Later on, Emilia changed her career and started working in Education as a Science Laboratory Technician, actively supporting the science department. Alongside her lab tech role, Emilia conducts workshops and STEM clubs from Nursery to year 11, promoting science among youngsters and delivering hands-on workshops in local primary schools. She is always eager to participate in STEM events and projects, both live and online. Emilia is on Instagram with the science page @emilia.science, boasting over 150k followers and collaborating with key UK brands like Breckland Scientific, Philip Harris, Gratnells, and the British Science Association.

In 2021, Emilia was awarded "Science Technician of the Year" by Gratnells. In 2022, she was highly commended as an outstanding STEM Ambassador by STEM Learning UK. In 2023, Emilia received the "Dukes Hero" award from the Dukes Education Group, recognizing her extraordinary contribution to their setting. She also became the Lab Technician and STEM Enrichment Duke Champion, facilitating all lab techs and staff involved in STEM enrichment within the Dukes Education Group. Emilia's mission is to inspire kids towards science careers while advocating for equality and diversity in the field.

STEM PROs Questions

Why do you do what you do? / Do you have a defining/ah-hah/ eureka moment where you knew what you wanted to do?

I am deeply passionate about science and education. My journey from working as a Medical Microbiologist to becoming a Science Lab technician and STEM Ambassador reflects my commitment to inspiring the next generation of scientists. Through workshops, STEM clubs, Science shows and my activity on social media I hope I can ignite curiosity and passion for science. Being able to share my passion for science is incredibly rewarding!

When I changed my career and started working with the children, I quickly realised how rewarding it was to see their curiosity and enthusiasm for learning. This realization inspired me to become a STEM Ambassador. I hope to inspire students and help them enjoy learning about science. Making a positive difference in their education means a lot to me.

What is something you wish you knew while in school?

I wish I had known that finding your path takes time, and it is ok to feel uncertain about the future. It is important not to be too hard on ourselves and to understand that failure is a natural part of learning. Instead of seeing setbacks as failures, they are opportunities to learn and grow. Everyone faces moments of discouragement, but it is crucial to persevere and never give up on your dreams. If I had known this earlier, I would have taken things more easily!

What is something you wish you did differently when you first started working?

Looking back, I wish I had learned to say 'no' more often. Setting boundaries and maintaining a healthy work-life balance is crucial to prevent burnout and ensure well-being.

If someone wanted to do what you do, what's the best piece of advice you'd share with them?

My advice is to prioritize continuous learning as knowledge is a powerful tool. Be passionate about your work and put in the effort needed to excel. With dedication and hard work, you can succeed in achieving your goals and making a meaningful impact in your field.

If you could definitively answer one unanswered science question, what would it be and why? / What unanswered scientific question keeps you up at night?

Even if this is not my field, I am fascinated about the undiscovered functions of the brain. Unlocking these functions has the potential to revolutionize our understanding of human cognition and behaviour, paving the way for innovative therapies and technologies.

What scientific discovery or event in history would you like to go back in time and witness?

I would choose to witness the discovery of penicillin by Alexander Fleming. This groundbreaking moment in history revolutionized medicine and saved countless lives by introducing the first antibiotic.

What's a misconception about your line of work?

A common misconception about science technicians is that we are only support staff for the science department. In reality, we have significant responsibilities beyond just support roles. From managing lab equipment to conducting experiments and ensuring safety protocols our contributions are crucial to the functioning of the Science department. Additionally, we contribute actively to the education of the students.

What's the best part of doing what you do?

The best part of my work is the joy and enthusiasm I see in children when they engage with science. Witnessing their happiness and eagerness to learn is a rewarding experience for me.

What is a necessary evil in your industry?

A necessary evil is the paperwork involved in managing laboratories resources, safety protocols, and risk assessment. While essential for ensuring smooth operations and compliance with regulations, these tasks can be time consuming and take away from direct engagement with students and hands-on activities.

Have you ever changed your line of work? If so, why and what was the change?

I transitioned from my initial career path in microbiology to working in education. Transitioning from my previous career presented its own set of challenges. Adapting to a new work environment, learning different protocols, policies and procedures required significant adjustment. However, my passion for inspiring young minds and promoting science education motivated me to overcome these challenges and thrive in my new role.

What purchase under $100 has improved your life: career or personal?

I recently bought a trolley for transporting resources during workshops. It has made the process of setting up and packing up for workshops much more efficient and convenient. Also, I can easily transport material, equipment and supplies from one location to another without having to make multiple trips or carry heavy loads.

How has social media benefited or hindered your career?

Social media, particularly Instagram, has provided me the chance to connect with a diverse network of individuals, including fellow professionals, educators, and science enthusiasts. Through Instagram, I have had the opportunity to build meaningful relationships and establish genuine friendships with individuals who share my passion for science and education. These connections have not only enriched my personal and professional life but have also opened doors to collaboration, mentorship, and opportunities for growth. Additionally, social media has allowed me to showcase my work, share insights, and engage with a wider audience, ultimately enhancing my visibility and impact within the STEM community. However, like any tool, social media also comes with its challenges, such as managing time and maintaining boundaries. However, the overall benefits of social media in connecting with others, building relationships, and advancing my career far outweigh any potential drawbacks.

STEM PRO

RUBY PATTERSON

RUBY PATTERSON: Former NASA Scientist, Geologist, Space Industry Consultant, Creator, Mastermind Group Leader, Entrepreneur, Podcast Host, Shih Tzu Mom

IG: *@rubythespacegeologist*
TT: *@rubythespacegeologist*
LI: *Ruby Patterson*

RUBY PATTERSON, a space geologist and PhD candidate, thrives as a space industry consultant at Astralytical where she helps space companies and start-ups solve their problems. Ruby previously worked as a Mars Geochemist at NASA's Johnson Space Center in Houston, Texas. While there, Ruby's work supported the Curiosity rover science team and Artemis astronaut tool development. Ruby finds immense joy in sharing her passion for space exploration with others. She dedicates her time to 1:1 mentoring aspiring scientists, nurturing the next generation's interest in planetary sciences, and inspiring them to reach for the stars. Ruby loves when people connect with her online, so don't be afraid to reach out to say hello.

STEM PROs Questions

Why do you do what you do? / Do you have a defining/ah-hah/ eureka moment where you knew what you wanted to do?

I'm a space geologist because it's one of the coolest jobs in the world. I love thinking about Deep Time, which operates the scale of millions to billions of years. I also see the tremendous benefit to society space exploration poses and love to be a part of something bigger than myself.

What is something you wish you knew while in school?

I wish I knew it was normal to feel confused and unsure of my path. It took me 4 years of an undergraduate degree, 2 years of a Master's degree, and the first 2 years of my Ph.D. to realize exactly what I wanted to do with my career. Never stop learning and trying new things!

What is something you wish you did differently when you first started working?

I wish I worked and did school part-time instead of full-time. I would have come out of school with more work experience and it would have opened my eyes to the way society works. Hint: it's very different from academia!

If someone wanted to do what you do, what's the best piece of advice you'd share with them?

Find a mentor who has walked the path you want to go down. Mentorship changed my life for the better, which is why I try to give back to folks coming up the ladder behind me.

If you could definitively answer one unanswered science question, what would it be and why? / What unanswered scientific question keeps you up at night?

Where the actual heck is Starship going to land on the lunar surface for Artemis III and IV? Is anywhere actually safe? This keeps me up at night.

What scientific discovery or event in history would you like to go back in time and witness?

I'd love to witness the first people talking through a telephone. Imagine how excited they must have been! It's probably similar to when I rode in an autonomous vehicle for the first time...

What's a misconception about your line of work?

That you have to be exceptionally smart to have a space career. Tenacity and hard work get you a lot further than natural intelligence ever will.

What's the best part of doing what you do?

I love sharing the joy of space with others. Yes, that's cheesy, but it's true. Getting a room full of kids excited about space exploration is one of the best feelings in the world.

What is a necessary evil in your industry?

Starship disturbing the upper layer of lunar regolith (AKA billions of years of geologic record) just to land a cargo ship and two people on the lunar surface. I support the Artemis program, it's just a tough pill to swallow that we are sending a rocket with a 100 ton payload capacity to the lunar surface repeatedly. For reference, 100 tons is the same weight as a blue whale.

Have you ever changed your line of work? If so, why and what was the change?

Yes! I left being a professional scientist to become a space industry consultant. I switched so I could make more money, better hours, and meet the folks on the start-up side of the industry. Working long hours in a science lab gets lonely sometimes!

What purchase under $100 has improved your life: career or personal?

Noise canceling headphones!!!

How has social media benefited or hindered your career?

Having a social media presence has given me more credibility and allowed me to connect with other young space professionals. It's benefited my career for sure.

STEM PRO
EDWARD GONZALES

EDDIE GONZALES: DEIA lead for Heliophysics at NASA Goddard, formerly NASA JPL

EDWARD GONZALES is an avid music lover and sneakerhead, Edward Gonzales has built a life walking the walk from street smart to business-savvy, first at one of Los Angeles's prestigious law firms, then to "student-whisperer" at NASA's Jet Propulsion Laboratory working with interns and underserved, underrepresented populations. Now, as NASA Goddard Space Flight Center's Program Manager for Culture, People and Equity for Heliophysics, he brings his unique blend of experience to one of NASA's most prestigious missions.

Edward has been honored by many NASA awards and recognition and has made it his personal mission to ensure that the agency's future workforce is more diverse and equitable than when he started his career there two decades ago. Having lost his father as a young teen, experienced police profiling and gang violence, he can relate to the struggles that many students face as they embark on their college and career journeys.

He is eager to share that all paths to NASA aren't linear and emphasizes the critical importance of creating a workforce pipeline that starts in school, leading to an exciting current project called "Permission to Dream," collaborating with Christopher Gardner (The Pursuit of Happyness) to present inspirational talks at one thousand highs schools across the United States. Edward is the consummate comeback kid—all setbacks set you up for a comeback.

STEM PROs Questions

Why do you do what you do? / Do you have a defining/ah-hah/ eureka moment where you knew what you wanted to do?

I want to make sure this arena is a safer place where one can be their authentic self and thrive, I had this moment of realization of "I can help others" while working at an International Law Firm before coming to NASA

What is something you wish you knew while in school?

That it's ok to make mistakes and to master handling hard better.

What is something you wish you did differently when you first started working?

Started my journey in helping others through Diversity a lot sooner.

If someone wanted to do what you do, what's the best piece of advice you'd share with them?

Network with others in your field. For example, I would google "Top professionals in the DEIA arena" and once I had a list of those folks, read any papers or articles they have written and then I would reach out to them via LinkedIn, with an ask of a 15 minute zoom to talk about their work, learn from those that have been there.

If you could definitively answer one unanswered science question, what would it be and why? / What unanswered scientific question keeps you up at night?

Is there life out there and where? This also keeps me up at night.

What scientific discovery or event in history would you like to go back in time and witness?

Well?? It would not be scientific, but I would go back to Martin Luther King's speech, be there live... We still have a long way to go.

What's a misconception about your line of work?

I have heard about Diversity fatigue as if it's a fad, not realizing how important it is.

What's the best part of doing what you do?

Seeing others have a voice, a safe place to work and being able to be their authentic self.

What is a necessary evil in your industry?

Always going to have nay-sayers

Have you ever changed your line of work? If so, why and what was the change?

Prior to working at NASA, I was part of Senior Staff at an International Law firm in Los Angeles, CA, I changed my line of work because I always wanted to work for NASA

What purchase under $100 has improved your life: career or personal?

A journal or notepad. Also the gel pen, love those.

How has social media benefited or hindered your career?

Mos Def benefited, it allows me to get my voice out there and lets others from wherever they are from know that I champion for them.

STEM PRO

NEHEMIAH MABRY

DR. NEHEMIAH MABRY: PhD Structural Engineering & Mechanics, MS & BS Civil Engineering, Educator, and Entrepreneur, Founder of @STEMedia, formerly NASA

IG: *@STEMedia, @NehemiahMabry*
TT: *@STEMedia*
X: *@STEMedia, @NehemiahMabry*
FB: *STEM.Media*
YT: *@stemedia*
Websites: *STEMedia.com, NehemiahMabry.com*

DR. NEHEMIAH MABRY is an Engineer, Educator and Entrepreneur based in Raleigh, NC. His 15+ years of engineering experience includes research at NASA, DOT Bridge Design and Analysis, and teaching as an Adjunct University Professor.

As a grad student, he founded STEMedia, a digital media company that provides educational and inspirational content to empower young professionals in STEM. Since its inception, Dr. Mabry has won national awards, thousands through public speaking and lectures, and partnered with several organizations, businesses and academic institutions.

STEM PROs Questions

Why do you do what you do? / Do you have a defining/ah-hah/ eureka moment where you knew what you wanted to do?

I am a structural engineer, STEM educator, and entrepreneur. I knew I wanted to do engineering when I had my first internship in high school at NASA. From there, my love for creative teaching came from being in an improv drama group in college, and in grad school, I realized that starting my own business was the best way to blend my passions.

What is something you wish you knew while in school?

I wish I knew (I mean really knew) that what other people think doesn't matter all that much. I mean, sure, you want your interviewer to get a good impression and your coach to see you as coachable, etc... But in the end, I wish I never allowed the fear of what others might think to delay the process of going after what I really wanted.

What is something you wish you did differently when you first started working?

I wish I had saved more money. To be honest, there was a financial valley that accompanied my initial jump into full-time entrepreneurship. Had I saved up more money, it would not have been so tough to trudge through that period—and believe me, it was tough.

If someone wanted to do what you do, what's the best piece of advice you'd share with them?

Two things... Number one, start now working towards it and don't wait. And number two, study the successes and failures of yourself and others along the way then make the appropriate adjustments with confidence and courage.

If you could definitively answer one unanswered science question, what would it be and why? / What unanswered scientific question keeps you up at night?

One science question that matters to me is what is the neuroscience behind the brain's ability to believe whatever it wants to, to the point that what it believes is true essentially becomes "true" and often seems to affect its external reality. Essentially, I am really intrigued by the science of faith.

What scientific discovery or event in history would you like to go back in time and witness?

I certainly would like to have witnessed the origin of sentient life here on earth.

What's a misconception about your line of work?

That I have all the answers and facts in my head; memorized. So much of what I do is read up on things that either remind me of what I have learned before or trigger a deeper understanding of something new based on principles I have learned before.

What's the best part of doing what you do?

I love inspiring people. Simplifying technical ideas and creative concepts and blending them in new ways to elevate and empower people is the best part of what I do.

What is a necessary evil in your industry?

As an entrepreneur, a necessary challenge is finding the money to fund the things that we want to do. So often, the things that are cool, and even needed at times, do not have the economic support at first and it takes tons of work to find the money before even starting the fun part. That is part of my job as CEO.

Have you ever changed your line of work? If so, why and what was the change?

I was once a full-time engineer and a part-time entrepreneur. I am now a full-time entrepreneur and a part-time engineer.

What purchase under $100 has improved your life: career or personal?

As a person of faith, I bought two books in college that greatly influenced my perspective towards my personal life and career, 1) "The Purpose Driven Life" by Rick Warren and 2) "How to Know God's Will for Your Life" by Morris L. Venden.

How has social media benefited or hindered your career?

Social media has definitely helped my career by connecting me with amazing people and bringing exposure and opportunities that I would not have had without it.

"Above all, don't fear difficult moments. The best comes from them."

–Rita Levi-Montalcini, neurologist and joint winner of the Nobel Prize in Physiology or Medicine for the discovery of nerve growth factor

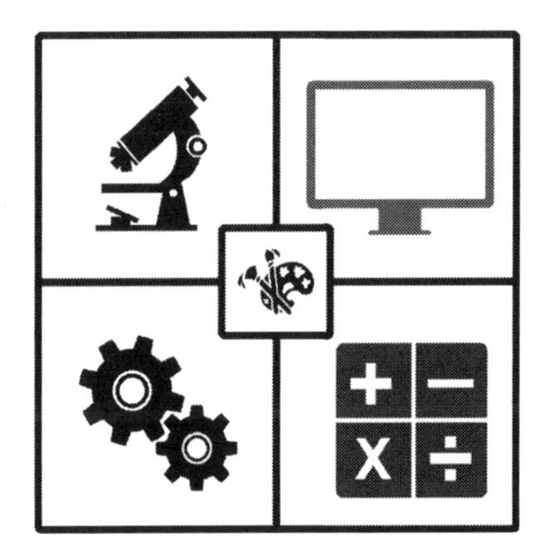

STEM PRO

DARIA PIDHORODETSKA

DARIA PIDHORODETSKA: PhD Candidate Earth and Planetary Sciences, BS Biology, formerly NASA

IG: *@planet.daria*
TT: *@planet.daria*
X: *@astrodaria*

DARIA PIDHORODETSKA is a fourth-year PhD student in the Department of Earth and Planetary Sciences at the University of California, Riverside. She completed her undergraduate studies in Biology in 2017, where she was the first student at her institution to conduct research in Astrobiology.

Fascinated by the search for life on other worlds, Daria went on to spend 2.5 years at NASA's Goddard Space Flight Center working on terrestrial exoplanet detection and characterization; through 1-D/3-D modeling and space-based observations, often using data from TESS (Transiting Exoplanet Survey Satellite).

Now, Daria works to expand these efforts by collecting data from the ground as an observer on the Keck Telescope while studying the atmospheres of planets in the outer habitable zone as a NASA FINESST Fellow. After graduation, Daria plans to work on future space-based telescopes that will be designed to detect and characterize Earth-like exoplanets.

STEM PROs Questions

Why do you do what you do? / Do you have a defining/ah-hah/ eureka moment where you knew what you wanted to do?

Science has always been captivating to me, but it wasn't until I was an undergrad that I knew I could work in astrobiology. There weren't any astronomy courses at my small university, so it wasn't until my junior year of college that I took a class called astrobiology and was opened up to this growing field where scientists were looking for life on other worlds. As soon as the first day of school was over, I knew this was something that would stick with me.

What is something you wish you knew while in school?

College is only the beginning. You still have so much time to figure out what you want to do, or to reinvent yourself if you're ready. So much of 1-12th grade is based on preparing you for college that it feels like the ending point in your life. Your life hasn't even started yet.

What is something you wish you did differently when you first started working?

I wish I took more initiative in understanding how much flexibility I had with my project goals and ambitions. As a new hire, I thought I needed to wait to be told what to do, and didn't understand just how much I was allowed to shape my own position. Don't be afraid to advocate for yourself!

If someone wanted to do what you do, what's the best piece of advice you'd share with them?

Get comfortable with rejection and failure. There is no linear path to working in science, especially astrobiology, but all paths typically experience hardships. Talk to someone that you admire in the field you want to work in and ask them about their former rejections. It will help you normalize this part of the journey.

If you could definitively answer one unanswered science question, what would it be and why? / What unanswered scientific question keeps you up at night?

Are we alone? This of course is the question that motivates my entire career journey. There are so many undiscovered worlds and we have never detected life on another planet. We discover planets beyond our imagination on a monthly basis, what does that mean for other life?

What scientific discovery or event in history would you like to go back in time and witness?

I'm extremely fascinated by the Pyramids of Giza and their unknown origin. It would be incredible to go back in time and see how they were built (and no, I don't think it was aliens).

What's a misconception about your line of work?

People think that astrophysicists are hiding secrets about UFOs and aliens. The truth is that scientists are absolutely terrible at keeping secrets, and if we knew they existed, it would not be long before everyone else knew too.

What's the best part of doing what you do?

There are never-ending questions about outer space, and the more we discover, the more we find ourselves asking. As a society, we have barely scraped the surface of space exploration, and the future is gleaming with opportunities to learn more about the Universe. I believe many of the most fascinating discoveries we'll make haven't even been considered yet. It's amazing to work in a field with such endless possibilities.

What is a necessary evil in your industry?

Math. You can't escape it, but you can minimize it. Learn to code and you won't have to worry so much about math, and count your blessings if you enjoy it, but don't feel bad if it's not your forte.

Have you ever changed your line of work? If so, why and what was the change?

Up until I took astrobiology my junior year of college, I planned to become a Physician's Assistant (PA) and work in a hospital. I don't think I ever saw working in astrobiology as attainable because I didn't grow up around it and I didn't see anyone like me working in planetary science. I was on a path that I thought was good for me, but I was on it blindly. Astrobiology opened up a passion inside me that I had never seen before. It was a no-brainer.

What purchase under $100 has improved your life: career or personal?

Apple air tags, I bought a bunch and put them in things I use daily, and I cannot recommend it enough. An electric tea kettle, any time I choose to take a yoga class, and a NYT online games subscription to destress.

How has social media benefited or hindered your career?

Social media has been a wonderful tool to connect me to others in my field as well as people of all ages who are excited to learn about space. I've been able to provide guidance to students all over the world and it has helped me understand what I want to do with my career. It's also given me incredible partnership opportunities that I am extremely grateful for!

STEM PRO

DR. SHAWNA PANDYA

DR. SHAWNA PANDYA: MD Medicine, MS Space Studies, Physician, Aquanaut, Explorer, Director—IIAS Space Medicine Group, Space Medicine and Austere Environment Researcher

IG: *@shawnapandya*
X: *@shawnapandya*
FB: *Shawna Pandya Official*
YT: *@ShawnaPandya*
LI: *@shawnapandya*
Website: *shawnapandya.com*

DR. SHAWNA PANDYA is a physician, aquanaut, scientist-astronaut candidate with the International Institute for Astronautical Sciences (IIAS), skydiver, pilot-in-training, VP Immersive Medicine with Luxsonic Technologies, and Director of IIAS' Space Medicine Group. She is also a Fellow of the Explorers' Club and Associate Fellow of the Aerospace Medicine Association. Dr. Pandya was on the first crew to test a commercial spacesuit in zero-gravity in 2015. She earned her aquanaut designation on the 2019 NEPTUNE (Nautical Experiments in Physiology, Technology and Underwater Exploration) mission, 2023 NEP2NE aquanaut mission. She served as Payload Crew and co-PI of the 2023 *IIAS-01* suborbital research flight, as well as a PI and/or co-I for Ax-2, Polaris Dawn and Blue Origin payloads.

Dr. Pandya's publications include a paper on medical guidelines for commercial suborbital spaceflight, and book chapters on space technologies that have benefitted terrestrial medicine, psychological resilience in long-duration spaceflight, and reproduction and sexuality in long-duration spaceflight. In June 2022, her extreme medicine work took her to Poland and Ukraine to work with persons affected by the Russian invasion of Ukraine.

In 2021, Dr. Pandya was granted an Honorary Fellowship in Extreme and Wilderness Medicine and named to the Canadian Women's Executive Network's Top 100 Most Powerful Women. Her work is permanently exhibited at the Ontario Science Center alongside Dr. Roberta Bondar, the first Canadian woman in space. In 2022, Dr. Pandya was awarded the Explorers' Club's "50 Explorers Changing the World," and in 2023, named to SustainabilityX Magazine's "Global 50 Women in Sustainability," for her work in Social Inclusion. In 2024, she was inducted as a full member of the International Astronautical Federation's Human Spaceflight Committee, became an aeromedical flight physician, and was nominated to the Women's Space Awards in the Medicine and Health category.. Her work has been profiled by Nature and the Royal Canadian Mint.

STEM PROs Questions

Why do you do what you do? / Do you have a defining/ah-hah/ eureka moment where you knew what you wanted to do?

I am passionate about exploring and pushing the limits, both as an individual, and of where we have been as a species. I like doing hard things. Growing up, there was no "aha" moment, but rather an entire childhood of being exposed to starry skies, and growing up in an era of the spaceflight of Canada's first female astronaut, Dr. Roberta Bondar. Watching Dr. Bondar, I knew that that's what I wanted to do!

What is something you wish you knew while in school?

There's no shame in not knowing—rather asking questions in order to gain an understanding is one of the best ways to learn!

What is something you wish you did differently when you first started working?

I wish I prioritized my own-well being as a medical resident. Whether I work 26 hours or 40 hours straight—the job will carry on without you, but you cannot perform at your best when your battery is only charged to 3%!

If someone wanted to do what you do, what's the best piece of advice you'd share with them?

If there is no path for what you want to do, make a path. Do it with like-minded, smart, accomplished people. Push the limits, and set ambitious goals. When you reach that goal, set an even more audacious goal and push even further! You'll be awestruck at where you end up.

If you could definitively answer one unanswered science question, what would it be and why? / What unanswered scientific question keeps you up at night?

Just ONE? Oh wow. In medicine—how do we preserve the brain against aging, neurodegenerative disease, and dementia? In performance—how do we best optimize cognitive focus, learning, and memory? In space medicine—how do we predict and enact countermeasures against the Spaceflight Neuro-Ocular Syndrome (SANS)? As you can see, I am a bit of a neuroscience nerd! Also...I want to know when (not if) we will find definitive proof of life elsewhere in the Universe.

What scientific discovery or event in history would you like to go back in time and witness?

There are too many, but the 1969 Apollo 11 Moon Landing definitely comes to mind!

What's a misconception about your line of work?

In medicine, there is a misconception that disease states in medicine are black and white, and that disease patterns and presentations always stick to the textbook definition. In reality, diagnosis and treatment is a lot more complex than that, and there is definitely a lot of sleuthing and nuance that goes into medicine. It's one of the parts of the job that I love most.

In space medicine, there is definitely an under-appreciation of how quickly the field is changing! The amount of knowledge, disciplines, studies and supporting technologies involved in supporting human health in space continues to burgeon at an exponential pace with each passing year.

What's the best part of doing what you do?

One of the best parts of what I do is working with teams of insanely brilliant, talented, and accomplished people.

What is a necessary evil in your industry?

I think wearing so many hats as a practicing physician, researcher, entrepreneur, advisor and explorer in space medicine and austere environment health mean that there is no such thing as a regular work day or a 9-5/Monday to Friday work pace. My days and weeks can stretch 16 hours a day, 7 days a week, and I live out of a suitcase most days of the month—but to be honest, I don't view it as an evil, I love it!

Have you ever changed your line of work? If so, why and what was the change?

Once upon a time, I was training to be a neurosurgeon. I eventually realized that that wasn't the field for me, for a variety of reasons, including that I was not able to focus on my space medicine and human spaceflight interests. I completed my residency in family medicine instead, and nowadays am a rural emergency physician while also working as an entrepreneur, researcher and advisor in space medicine.

What purchase under $100 has improved your life: career or personal?

A skipping rope! Physical fitness is extremely important to me, so I bring my workout gear and a skipping rope everywhere I go so that I can stay active, even when I don't have access to a gym.

How has social media benefited or hindered your career?

Social media has mostly benefited my career. I have been lucky in that there is a lot of interest in support in my journey, ideas and adventures. I have been able to make meaningful contributions to important conversations in space, medicine and key social issues through my online presence. I have connected with amazing space professionals, and have also come into some amazing opportunities via social media.

STEM PRO

ZAHRA RONIZI

ZAHRA RONIZI: Analog Astronaut, Former NASA Intern, Space Medicine Researcher

IG: *@themarsgal*

ZAHRA RONIZI is an undergraduate student at Harvard University pursuing biomedical engineering with a focus on space medicine. She is an analog astronaut, student researcher, and STEMinist devoted to female advocacy in STEM. She strives to help facilitate human exploration of deep space as a future flight surgeon and bioastronautics researcher.

Zahra is a former intern at NASA's SEES program, where she spearheaded the development of a conceptual rover mission on the Mars Rover Resource Utilization team. She serves as the founder of the Odyssey Scholarship Program, an initiative through which underrepresented students across the world attend STEM summits and receive mentorship on pursuing careers in the space industry. She is a published researcher with the American Geophysical Union and has recently been featured on Nature for her work as the crew biologist on the Astroland Interplanetary Agency's first all-female analog mission inside a cave. She has also served as a mission specialist on analog missions in Iceland and at sea. Apart from her galactic endeavors, Zahra loves teaching karate as a sensei, scuba diving, and traveling!

STEM PROs Questions

Why do you do what you do? / Do you have a defining/ah-hah/ eureka moment where you knew what you wanted to do?

As an aspiring flight surgeon and astronaut, I'm pursuing something that's a million worlds beyond the goal of just one person; rather, it's the culmination of all of humanity. In the space industry, I'm most drawn to the fact that space exploration connects people of all nationalities and walks of life with the shared passion for discovery. In this regard, humanity stands united.

What is something you wish you knew while in school?

I'm still in school, so every day is a new learning lesson for me! I wish I had understood the importance of maintaining a "balance" early on. Oftentimes, students try to juggle a million different opportunities to boost their resumes, forgetting to nourish their personal needs. Put forth your maximum effort in furthering yourself academically and professionally, but always make quality sleep, time with loved ones, and good self-care habits your top priorities. Ultimately, a healthy body and mind will be the fuel you need to chase your ambitions!

What is something you wish you did differently when you first started working?

I truly believe that all the ups and downs of life are necessary to shape a person (and a professional) into who they are. With that being said, I wish I would've reached out to professionals in my field of interest early on to learn about the paths they took to reach their goals. Having examples to follow would've definitely been less stressful than trying to engineer a completely new path for myself!

If someone wanted to do what you do, what's the best piece of advice you'd share with them?

Pursuing a career in space is never a linear path, so expect the unexpected. While the academic route to attaining a certain career may be clearly laid out, always be open to taking on different opportunities while keeping your end goal in sight. All these unique experiences will provide you with new skills and perspectives that will propel you towards your endeavors!

If you could definitively answer one unanswered science question, what would it be and why? / What unanswered scientific question keeps you up at night?

As David Bowie once melodically asked, "Is there life on Mars?" Since the 1970s, NASA has sent an army of satellites and landers to visit the red planet and search for clues. Since then, countless signs have emerged that Mars may still host microbial life. This is the question that inspired several of my research projects on analog missions and continues to drive my curiosity today.

What scientific discovery or event in history would you like to go back in time and witness?

The Apollo 11 moon landing, hands down. One of my favorite parts of each space mission is watching the live reactions of the team in mission control after they've received the first update from the spacecraft. It's the most heartwarming moment to see all the scientists and directors jump out of their seats and embrace each other in immense joy after their years of hard work has paid off. I would give anything to be in that room when Neil Armstrong told the world, "the Eagle has landed."

What's a misconception about your line of work?

One of the biggest misconceptions I hear is that space travel only serves the purpose of exploration. Beyond the desire to venture through new worlds, there has been an entire universe of research and discoveries made in space that have profoundly benefited our world back home. The International Space Station, the world's only long-duration research laboratory in space, has dedicated over 20 years to driving innovative solutions on Earth. Research on Alzheimers' Disease, water purification systems, weather-tracking programs, fluid physics...you name it!

What's the best part of doing what you do?

Witnessing the limitless diversity. One of the most mind-boggling aspects of the space industry is the incredibly wide range of people you meet, both culturally and professionally. Space is a global venture and unites the Earth under the goal of exploration--through my space dream, I've had the opportunity to connect with individuals from across the world and learn about all their beautifully unique cultures. It's also so impressive to work with different professions that are connected to the space industry. Beyond the engineers and scientists, I've worked with space chefs, astronaut fitness trainers, and space art designers!

What is a necessary evil in your industry?

The stressful urge to beat the clock.

Have you ever changed your line of work? If so, why and what was the change?

Not as of yet.

What purchase under $100 has improved your life: career or personal?

My full-coverage hijab undercap. From blazers to flight suits, styling professional clothing with the hijab often becomes a struggle in terms of preserving my modesty. Investing in durable undercaps from Muslim-owned small businesses has helped me rock my outfits without constantly worrying about my hijab slipping throughout the day. Whether it's rock-climbing on an analog mission or presenting at a conference, I can now focus entirely on the task at hand while proudly representing my religion as a hijabi!

How has social media benefited or hindered your career?

Social media has always been a blessing for my career. Starting my own science communication page, I was able to connect with so many other aspiring astronauts at a young age and join a supportive community of people passionate about space travel. As I grew in experience and knowledge, I was able to share my space pursuits with my platform, proving to others that they should never let their age stop them from reaching their goals. I look forward to continuing to share my journey to space on social media in the years to come!

STEM PRO
LEE GIAT

LEE GIAT: Filmmaker & Pilot

IG: *@astrodirector*
YT: *@flyingostrich*
Websites: *flyingostrichmedia.com, passageflight.org*

LEE GIAT is an aerospace industry filmmaker and pilot. His company, Flying Ostrich Media (@flyingostrichmedia), is working to create the first feature film on the moon.

He also founded a project called PASSAGE (@passageflight), which delivered over $75,000 of science education resources to underprivileged kids across the Caribbean and Latin America.

STEM PROs Questions

Why do you do what you do? / Do you have a defining/ah-hah/ eureka moment where you knew what you wanted to do?

To give some credit to the astrology people (I know, ew): I do feel like a Gemini. There are definitely two Lee Giat's in my brain. For as long as I can remember, fate had decided I would become both a filmmaker and pilot. In 2005 at the age of seven, I made my first short film using my mom's old camera, and edited on Windows Movie Maker on a dinosaur PC. Coincidentally, my first logged flight happened that very same year when my dad took me on my very first flight lesson.

What is something you wish you knew while in school?

I really have no regrets about how I carried out my school days. I had a "C's to get degrees" mentality from an early age, and just wanted to make the most of my life as a child and young adult. In high school, I had a blast just hanging out with my friends. We rode bikes and flew planes, and it's one of the most cherished times of my life. In college, I kept that same mentality. I knew that networking would be my biggest asset toward success, and practicing my craft outside the classroom. Look, many people love the classroom environment. I was most definitely not one of them. Know yourself, know how you learn, and don't forget to have fun. Right now is the youngest you'll ever be.

What is something you wish you did differently when you first started working?

I wish I had more jobs early on. Today, I have the privilege of being my own boss, and running a company surrounded by incredible people. This started fairly quickly after graduating college, and I wish I had more opportunities to work for others and gain insight on how good businesses function.

If someone wanted to do what you do, what's the best piece of advice you'd share with them?

Something people tend to forget is that today's world moves at a faster pace than ever before. It can be intimidating, and social media makes it feel like there are millions of people exactly like you. I certainly feel that way, and it's very discouraging as a creative. My Instagram feed continues to be bombarded with posts of young men my age that are both filmmakers and pilots. Ignore the noise, do things your own way, and bust your ass.

If you could definitively answer one unanswered science question, what would it be and why? / What unanswered scientific question keeps you up at night?

Who, what, where, when, why, and how. The six fundamental questions of our existence that, if answered, would give us a complete understanding of our being here. Even if only one of those were answered, that would put humanity thousands of years ahead in solving our greatest scientific mystery.

What scientific discovery or event in history would you like to go back in time and witness?

Apollo 11. It was the most-viewed televised event in human history. Even 50+ years later, it's still hard to believe that our tiny species set foot on the moon. Even though we're destined to do it again, I would've loved to witness the first time we walked on another world.

What's a misconception about your line of work?

That it's always fun. Spoiler alert, it's not! Look, I love the fact that I get to wear pajamas to work, write scripts on oversized bean bags, and play music all day in our studio to get the creative juices flowing. But with that comes lonely editing all-nighters, sweaty Florida summer shoots, lots of heavy lifting, writer's block, and weeks away from home. It may sound fun and romantic at first, but sometimes you miss certainty. The routine of a normal life.

What's the best part of doing what you do?

Wearing whatever I want. My work uniform is a hoodie with an ostrich on it.

What is a necessary evil in your industry?

Self-promotion. Unfortunately, people don't just "know" you exist. Like in any industry, dedicating time toward sales is super important if you want others to take you seriously. With our Flying Ostrich Media hoodies, local outreach events, and video content; people begin to recognize our logo, faces, and work!

Have you ever changed your line of work? If so, why and what was the change?

Nope!

What purchase under $100 has improved your life: career or personal?

My Perry Ellis backpack. I've had it for nearly 14 years now, and it's traveled with me to 12 countries without fail. Not sponsored, but I really love this brand!

How has social media benefited or hindered your career?

These days, I am only on social media because of my company and charity projects. It's not really a personal outlet for me as it used to be. Many platforms have become oversaturated with influencers, ads, and misinformation. However, I love that social media has the power to connect the right people with the potential to change the world.

"Most people say that it is the intellect which makes a great scientist. They are wrong: it is character."

–Albert Einstein, theoretical physicist known for developing the theory of relativity and winner of the Nobel Prize in Physics

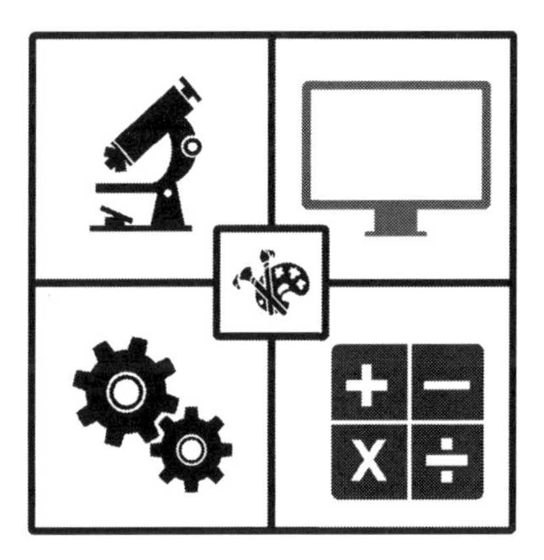

STEM PRO

CHLOE OSORIO

CHLOE OSORIO: Professional Civil Engineer (PE), STEM Advocate

IG: *@chloetheengineer*
TT: *@chloetheengineer*

CHLOE OSORIO is a civil engineer in the power industry in Los Angeles, CA. Her mission is to show that women can "be both" and to break stereotypes of women in STEM. Chloe is the founder of the #stompingoutstereotypes movement, which showcases that women in STEM are multifaceted, whatever that looks like to them. Her platform shows her life as an engineer, real estate investor, entrepreneur, dog mom, and everything else in between. Her goal is to show that engineers are all unique and that is a good thing!

STEM PROs Questions

Why do you do what you do? / Do you have a defining/ah-hah/ eureka moment where you knew what you wanted to do?

Growing up I knew I wanted to do something that helped people. Well, whenever you tell people that, they assume you will be a doctor or nurse. However, I am TERRIFIED of blood and needles and medical everything, so healthcare was a hard no for me. In high school I explored architecture and even took college level courses thinking that would be the path I would take. However, as I learned more about civil engineering, I realized I could combine my love of architecture and design with my love of math and science and it seemed to be the perfect fit. I also soon realized that civil engineering is a field that helps entire communities at a time. While people in healthcare are absolutely amazing, they help one patient at a time. What I realized about civil engineering is that with one project you can help an entire community all at once, and that is how I knew I would be passionate about it.

What is something you wish you knew while in school?

One thing I wish I knew in school was that how you do in each individual class does not reflect how you will do in your career. School is just preparing you to think like an engineer, but your experience in the workforce is what truly matters. I wish I knew to not take it all sooooo seriously and to be a little less hard on myself.

What is something you wish you did differently when you first started working?

When I first started working, I wish I asked more questions right away, and did not feel that I had to prove myself. Now, I tell all the young engineers I train and mentor to ask questions and remind them that we will not think less of them for not knowing an answer. When you are first out of school, it is not expected you know everything. However, I felt like I would not be taken seriously unless I could prove myself, and asking questions would show weakness.

If someone wanted to do what you do, what's the best piece of advice you'd share with them?

My best advice for anyone who wants to be a civil engineer is to learn from everyone and to be willing to be collaborative. Sometimes engineers think that they have to call all the shots, but really the folks that are on the construction team or in operations often know as much if not more than us about how everything should be designed and constructed. Listen to everyone's input, learn from everyone you work with, and always remember that no matter how many degrees you have or how many years of experience, there is always more to learn.

If you could definitively answer one unanswered science question, what would it be and why? / What unanswered scientific question keeps you up at night?

I would love to truly truly know how the Ancient Egyptian Pyramids were really built. Even though there are many theories and studies, we probably will never truly know how that feat of engineering was completed!

What scientific discovery or event in history would you like to go back in time and witness?

I would love to go back in time and witness the construction of the Brooklyn Bridge. I would absolutely love to meet Emily Roebling and see how she was able to take charge of the project and accomplish such an amazing feat of engineering during a time when women were not seen as being intelligent enough to be engineers!

What's a misconception about your line of work?

One misconception about civil engineering is that we basically just work on buildings and bridges. However, civil engineers are responsible for pretty much all of our critical infrastructure that we need to function as a society. Some other things that civil engineers work on is drainage systems, sewer systems, roads, power systems, drinking water supplies, communication systems, parks and recreation facilities, waste disposal, and so so much more. If you step outside and look anywhere, you will see infrastructure that a civil engineer was a part of.

What's the best part of doing what you do?

The best part of doing what I do is knowing that I am truly making a difference in my community. I specifically work in the power industry and work on power distribution projects, and it is amazing to know that I have a direct part in making sure that my community has access to safe and reliable power.

What is a necessary evil in your industry?

A necessary evil in my industry is the bias that sometimes accompanies being a woman in engineering and construction. The power industry, and the construction industry as a whole, is extremely male dominated still to this day, and although 95% of the people I work with are amazingly supportive, there are some that still do not think women belong in the industry. I work hard to break those stereotypes and be the best engineer I can be, and I hope that one day this will no longer be a necessary evil in my field of work.

Have you ever changed your line of work? If so, why and what was the change?

I have only changed my line of work once, and very early in my career. I started my career as a civil engineer in the transportation industry, and although I learned so much and loved the company I worked for, I just knew I was not passionate about the projects I was working on and the designs I was working on were not as interesting to me. I received an opportunity to work in the power industry and I took it, and have not looked back since. Power and electricity fascinate me and I am extremely passionate about the projects I get to work on every single day.

What purchase under $100 has improved your life: career or personal?

This question took some thought! One thing I recently purchased for just under $100 is custom inserts for my running shoes, and that has actually improved my life both personally and professionally. The reason is, running has become an outlet and stress reliever for me, but I have been having a lot of pain that was hindering my ability to run longer distances and not allowing me to have the outlet. However, after purchasing the custom inserts I am able to enjoy running again which has helped me in more ways than one!

How has social media benefited or hindered your career?

Social media has been amazing for me in my career. It not only has allowed me to create a platform and share my life as a civil engineer, but it has allowed me to create a community and connect with so many people around the world. I have had opportunities both professionally and personally come from my presence on social media, and I have made connections that have turned into amazing friendships. One of my favorite things that has come from social media is the #stompingoutstereotypes movement that came simply from posting one photo of a steel toed boot and a heel. That hashtag has brought women from all over the world together for one common mission... to break stereotypes of women in STEM! But, most importantly, social media has allowed me to show the next generation that they can be an engineer too, no matter who they are or where they come from.

STEM PRO

MORGAN A. IRONS

MORGAN A. IRONS: Soil Scientist, Astroecologist, Cornell University Soil and Crop Sciences PhD candidate, Co-Founder & Chief Science Officer at Deep Space Ecology, Science Communicator, BS in Environmental Science & Policy and Biology from Duke University

IG: *@astroecologist*
X: *@Astroeco_Morgan*
FB: *Morgan Irons, Scientist*

MORGAN A. IRONS (they/them) is a soil scientist and astroecologist working at the intersection of the agriculture and space industries. At the time of this writing, Morgan is completing a Ph.D. in Soil and Crop Sciences at Cornell University. Their Ph.D. research focuses on microbial and organo-mineral interactions in soil systems experiencing different gravity conditions. Morgan is also a Founder and Chief Science Officer of Deep Space Ecology, Inc., a startup working on the challenges of food insecurity and human sustainability in the deep spaces of Earth, the Moon, Mars, and beyond.

Their academic and industry research over the last nine years has led to the first natural Earth soil and biochar media experiments on the International Space Station and with Blue Origin, the development of the Pancosmorio Theory for human space settlement and the Terraform Sustainability Assessment Framework for bioregenerative life support systems, the creation of the quasi-closed agroecological system model for extreme environment habitation, and soil management methods for regolith conditions.

Morgan's dedication to mentorship and teaching is evident by their commitment to communication through social media and 40+ academic and industry talks, their outreach work, and the research independent study projects they help develop and mentor for students wanting to enter the nascent fields of astroecology and terraform engineering.

Morgan holds various appointments, such as the position of an expert judge for NASA and CSA's Deep Space Food Challenge, a Cornell Atkinson Center for Sustainability GR Fellow, a National Science Foundation GR Fellow, a Carl Sagan Institute Fellow, a Norfolk Institute Fellow, the 2019 Ken Souza Memorial Spaceflight Award recipient, a 2017 Brooke Owens Fellow, and a 2017 Motherboard VICE Media Human of the Year recipient. Morgan received their Bachelor of Science degrees in Environmental Science & Policy and Biology from Duke University with a minor in chemistry.

STEM PROs Questions

Why do you do what you do? / Do you have a defining/ah-hah/ eureka moment where you knew what you wanted to do?

No matter where people might find themselves, everyone deserves health, safety, security, and the freedom to thrive and embrace their culture and traditions. Food security is a major part of communities being able to experience such aspects of life. We all must eat to be healthy and to live and yet not all of us have the luck or privilege to do so consistently or with nutrition prioritized. With the impacts of climate change and the consequences of intensive, conventional agriculture, many communities are losing access to localized food systems as well as environments that once supported plants endemic to their traditions and food culture. I work at the intersection of the agricultural industry and the space industry as a researcher and entrepreneur because I want to contribute to a present and a future where people can have sustainable and resilient agricultural systems that support their communities and allow them to thrive, no matter where they call home.

What is something you wish you knew while in school?

As I write, I am finishing up the last year and a half of my Ph.D. program, so I am technically still in school. What I wish I knew at the beginning of my school journey that I am benefiting from now was how to build "systems" for my habits. As "Atomic Habits" author James Clear says in his book, "Goals are about the results you want to achieve. Systems are about the processes that lead to those results" (page 23). From kindergarten to high school graduation, the school system I went through had strict and detailed systems in place that, if utilized, could lead to a student being successful (with the caveat that school life is not a vacuum and will be impacted by society, family dynamics, changing health, etc.). I developed into a student who utilized these systems and had parents who made sure I utilized these systems. This is not inherently a bad thing as I still had some freedom and opportunity to develop my own study habits and systems for performing as a student. However, when I went off to college, the academic experience became less structured in the sense that I was more independent and responsible for my daily actions and living environment. I had many adjustments to make, and I ultimately succeeded in my undergraduate studies with the help of mentors, friends, and the Duke University Academic Resource Center; however, looking back I was stressed (sometimes distressed), anxious, and did not have healthy boundaries with my academics. My life was my academics and activities. Always working. Always busy. Little to no 'true' downtime. This still might seem normal as college is a busy and stressful time. However, when I went off to graduate school, I had a harsh reality check.

Graduate school is an independent and sometimes isolating experience that requires focus, commitment, and much mental and emotional labor. I am not trying to scare anyone from going to graduate school. However, I hope that the lessons I have learned will provide a prospective graduate student with tools and guidance to have an easier transition into graduate school or an opportunity to regain perspective if already in graduate school.

The "systems" or processes that I had developed to be successful from kindergarten to receiving my Bachelor of Science degrees needed to be reevaluated and reconstructed to adjust to the graduate school experience. I did not realize this. It seems obvious now due to hindsight bias, but I felt that since I had been successful in my undergraduate years and research experiences, I was ready and prepared for graduate school. I did not realize that my systems at the time were not constructed to handle what I was about to go through. Three years into my Ph.D., I burned out from overworking myself and from the emotional distress of being isolated and anxious during the COVID-19 pandemic. My burn out got to the point where I might have dropped out of my Ph.D. if I had not done anything to change my situation.

The next three years, even today as I write this, I have been in weekly therapy to recover from my burn out and to work through the challenges of graduate school. You could say the first two years of burn out became my "learning" and "experimenting" years (to use Jay Shetty's dynamic approaches terminology from his 5 January 2024 On Purpose podcast episode), where I was learning how to re-engage with my work and trying to figure out how to rebuild what I did not realize at the time were my shattered systems—the very processes that were needed to help me complete my goals. Even as I started implementing the strategies and tools that my therapist recommended, and "healthy habits" I spent months experimenting with, I still felt adrift and could not see myself making progress on my Ph.D., even when I was. The major turning point in this experience, where I started to transition from "learning" and "experimenting" to having "performance" days, weeks, and months was after I read "The Miracle Morning" by Hal Elrod and "Atomic Habits" by James Clear. These two books are where I learned about the science and application of habits and the idea of systems. You could say that these books provided me with an eureka moment where all the tools, strategies, and habits I had been learning and experimenting with over the last two years finally came together into a comprehensive system that made sense and provided a mind shifting clarity for me. It was knowledge that I personally needed to finally overcome what felt like chronic impacts from my burn out. I am happy to say that I am now going strong and on schedule to finish my Ph.D.

I encourage anyone, no matter where you are in schooling or in your journey outside of schooling to read "Atomic Habits" and "The Miracle Morning" as I wish I had gained an understanding of habits and systems earlier in my schooling.

What is something you wish you did differently when you first started working?

Understanding that I can say and sometimes should say "no" to taking on different tasks, positions, projects, etc. The skill to say no comes with understanding my boundaries, my priorities, my job description and associated compensation, and my mental, emotional, and physical capacities. It is easy to train ourselves to say "yes" to everything as it might have, at some point, been great to build our resumes, to gain experience, to help friends or colleagues, to not upset someone, to make us feel good or needed... However, there comes a point when saying yes can take away the time and energy needed for the things we need to prioritize. Activities and positions that I would have said yes to in the past are no longer priorities as my commitments have changed with my life and career. When I do become unsure on whether to accept or decline a potential commitment, I talk with my mentors and advisors to gain their perspectives before making my decision.

If someone wanted to do what you do, what's the best piece of advice you'd share with them?

Not every school or location will have classes, research, or internships that align exactly with what you want to do. My advice to anyone seeking to do what I do is to create their own opportunities for research when none are available. One way in which someone might create their own opportunity is by performing a research independent study with teachers, faculty, professionals, or graduate student mentor(s) who are working in a field within or related to your interest. A research independent study is a process in which an individual, with the guidance of a mentor, delves into the literature and research of a topic of interest. The study could then progress from a literature review stage to experimental design, research proposal, and eventual experimentation, analysis, and paper publication. The key starting steps to begin the research independent study process are (1) to have enough knowledge to articulate the research topic or idea that you are interested in exploring; (2) to understand your initial goals for performing the independent study and long-term goals that could come from performing the study; and (3) have a list of potential research mentors whose research aligns with your specific topic or overarching field of study. It is important to note that some academic institutions might have research independent study programs that are available for course credit.

To provide a personal example, I realized I wanted to study closed ecological systems for space habitation as a sophomore at Duke University. At that time, no faculty performed closed ecological system research or space habitation research. However, I was surrounded by environmental scientists who were experts in earth-based ecological systems and habitation. The three mentors I found were professors who were willing to put a "space-twist" to their thinking and go on this independent research journey with me. Thus, I followed my first three steps: (1) The research topic I wanted to explore was understanding how ecological theories and principles could contribute to or solve open questions in the development of closed ecological systems for space habitation. My initial exploration into the research revealed an engineer-heavy perspective, and I wanted to bring forward an ecological perspective while delving further into the research. (2) My initial goals for performing a research independent study were to provide myself with the space and time to explore closed ecological system research, to gain mentored experience in the literature review process, and to write a paper detailing my findings from the review. My longer-term goals at the time were to have this initial study lead into designing and proposing a laboratory experiment. (3) I created a list of environmental scientists working at Duke who were performing research that would be relevant to my topic and wrote professional emails detailing out who I was, why I was emailing them (to ask for mentorship through a research independent study), explaining my topic idea and goals, how the topic aligned with their (the professor's) research interests, and asking if they would like to put a "space-twist" on their understanding of ecological systems. My research journey that led into the career that I have today started with a research independent study project.

If you could definitively answer one unanswered science question, what would it be and why? / What unanswered scientific question keeps you up at night?

The question that keeps me up at night is how we can [re]vitalize soil after it has been degraded—a question relevant for both Earth and space systems. I consider it a grand challenge as there are many questions and hypotheses that can be asked and tested. It is also an interdisciplinary and global question that requires communication and teamwork between people from different fields, industries, and countries to gather data and find broad and site-specific solutions.

What scientific discovery or event in history would you like to go back in time and witness?

I would like to witness the scientific discoveries that have not been recorded in history or have been miscredited due to biases and restrictions of the times. To know the truth of history, facing its harsh realities head on, makes the actions done by people restricted by their societies, who sought to discover and change the world, that much more powerful.

What's a misconception about your line of work?

I work at the intersection of the agriculture and space industries. More specifically, I work as a soil scientist and astroecologist. A common misconception about my work is that I do not solve or care for challenges on Earth because I work in the space industry. The fascinating reality of my work is that Earth and space food security and sustainability challenges are quite similar, so I can solve both at the same time. As a scientist of the environment, I understand how challenging quantifiably sustainable habitation is to achieve—the Earth took billions of years to evolve into its current state. This perspective allows me to understand the reality of needing to conserve our Earth while also making sure people who go out into space do not repeat the same environmental and social mistakes we have made here on Earth.

What's the best part of doing what you do?

My academic and professional careers have given me the opportunity to meet so many people from around the world. If I can learn anything from my career it is the importance of widening my world view and asking questions. My life and science are further enriched by the people I have met and will continue to meet.

What is a necessary evil in your industry?

I believe that to accept "necessary evils" is to perpetuate the activities that are considered "evil" instead of putting pressure on industries to change and create solutions. I acknowledge that with my identity and circumstances this can come off as a privileged and contradictory statement as the society I live in and participate in benefits from these "necessary evils" in the agriculture industry and space industry. It is also not my right to judge, i.e., a farmer who uses pesticides to lower the risk of them losing their crops—pesticides being considered a "necessary evil" for many people and organizations in the agriculture industry. What I can do is use my career to research and find alternative solutions to these "necessary evils" and work with people to find accessible ways to implement such solutions. I can also be mindful as a consumer when I am making decisions and communicate my knowledge to people who want/need it.

Have you ever changed your line of work? If so, why and what was the change?

As many students do in their undergraduate studies, I changed my major and eventually the professional path I took after my bachelors. I originally planned to major in biophysics, complete the pre-medicine track, and go off to medical school to pursue a career as an orthopedic surgeon. I did complete the pre-medicine track as I was still considering medical school until graduation; however, I ended up double majoring in Environmental Science & Policy and Biology with an ecology focus and a chemistry minor. I then went off to Cornell University for a Ph.D. in Soil and Crop Sciences. Why and when did everything change? It seems like a drastic change to make.

After two years and summers of biophysics studies, I had to come to terms with the cognitive dissonance I felt towards what I thought would be my major. I had decided back in high school that I would major in a biomedical or biophysics field as I believed that would provide me with the best foundation for medical school. It was also the major I told friends, family, and teachers I would pursue in college. The expectations I set for myself, the expectations I believed others had for me, and the reality of my experience trying to meet those expectations did not match up, leading to the cognitive dissonance and anxiety I felt while in college. The way I was able to settle my anxiety and dissonance was to talk with friends and family about my fears, challenges, and concerns with the path on which I found myself. I now know a therapist would have been a wonderful resource for me at this time, but I did not have the insight into mental health care that I do now. Thus, I am thankful for the support I was provided by friends and family members.

My friends and family helped me realize the obvious truth that the field of study about which I was passionate, found excitement in studying, and had quite a bit of experience in was environmental science. It was a childhood dream of mine to understand how and why our Earth has such diverse ecosystems. I also wanted to know if such ecosystems were found on other planets. I had been involved in many environmental initiatives through the years, and even at Duke, within the first week of being on campus, I became the Tree Campus Ambassador with the grounds and facilities department. It is quite fascinating how passions and dreams can be smothered or hidden in plain sight simply by believing we must meet certain expectations, whether placed on us by ourselves or those around us. Anyone who knew me would have wondered why I was pursuing biophysics instead of environmental science. I can vividly remember the conversations and expressions on my friends' faces when I asked whether I should switch to environmental science and biology. To sum up, it was a "no duh" moment.

The experience that eventually sealed the deal for me to switch my major at the end of my sophomore year and eventually decide on a Ph.D. instead of medical school was a series of research independent studies I performed with three Duke Nicholas School of the Environment professors as my mentors. The research independent studies I performed culminated in an experiment in which I studied the development of biogeochemical mechanisms in sterilized Mars regolith simulant to drive primary succession in the establishment of a closed ecological system on Mars. Pretty much, how do we develop ecological systems from scratch in extreme environments. The research I performed, the opportunities it brought to my academic career, the creation of my company Deep Space Ecology, and my growing fascination with scientific research led me to make the critical decision to pursue a Ph.D. instead of medical school.

Even though I do not regret my decision to switch the trajectory of my career, I remember how difficult it was to make such a decision. I believed since middle school that I would become an orthopedic surgeon. (You can find my story in several podcasts and YouTube video interviews if you want the details behind my original decision to pursue medical school.) Even as I had success with switching my major and creating my research career, I still completed the pre-medicine track, performed hundreds of hours of hospital volunteering and medical training, and completed a medical fellowship because I truly believed that by the end of my undergraduate studies, I would choose medical school. When it came down to the decision, I had two very different paths that would lead to great opportunities and fulfilling careers. It was a hard decision to make.

I chose to pursue a Ph.D. because I realized I wanted to continue in scientific research. I wanted to continue asking questions and designing experiments. I realized that my research could not only impact space habitation but also had direct applications to agricultural innovation and food security on Earth. Through working in agricultural research, I could also bring in my medical experiences as human health is interconnected with food security and the sustainability of our agricultural systems. I would not become an orthopedic surgeon, but as a Ph.D. and eventually a career researcher, I had the opportunity to develop and provide knowledge and tools and innovations that could empower people to improve their health and communities.

What purchase under $100 has improved your life: career or personal?

At the current time, five books have improved my career and personal life, and combined in their cheapest format, are under $100—prices vary depending on book format. The books are "The Miracle Morning (Updated and Expanded Edition)" by Hal Elrod; "Atomic Habits" by James Clear; "Think Like A Monk" by Jay Shetty; "Braiding Sweetgrass" by Robin Wall Kimmerer; and "Start with Why" by Simon Sinek.

How has social media benefited or hindered your career?

I started using social media to communicate with friends and family, something everyone does. I also treated it as a personal journal, a place for me to write about my day and record events through image and video archives. I believe my habit of using social media as a personal journal led me into a science communication journey that I had no idea would benefit my career. I started communicating my research on social media when I was a sophomore at Duke University. I wanted to show the skills and knowledge I was gaining in my research experiences and through my independent studies. Slowly, my social media accounts became all about my research and academic journey. This, unknowing to me at the time, created a personal brand that people would come to know and associate with me. I became the "Astroecologist", a young career researcher who was performing research (space agriculture research) many people in the public had never seen or heard of before. I was not shy about sharing every step of my research process and even became known for my scientific methodology videos. I came to realize that by being open about my research, I was getting engaged beyond my friends and family. Teachers and organizations I had no prior association with, such as The Mars Generation, reached out to me to ask for my involvement with their educational activities. I received a leadership position in The Mars Generation through engaging with the organization after we communicated on social media. The opportunities I received did not stop as my science communication went further than I could have imagined. I received talk invites from US and international conferences and university departments. I have shared my research with thousands of K-12 students. I have helped students start their research and even publish it. I have received awards, such as VICE Media's Human of the Year Award for 2017. I am also a subject matter expert judge for NASA and CSA's Deep Space Food Challenge...

The Roman philosopher Seneca is attributed with saying, "Luck is what happens when preparation meets opportunity." Social media was a tool that gave me the location and time for me to prepare and communicate my preparedness with people and organizations that then could then offer opportunities. I was lucky, and it took a lot of work and time. I still use social media as a tool to communicate my science, and I will continue to do so.

STEM PRO

DR. MALIKA GRAYSON

DR. MALIKA GRAYSON: PhD Mechanical Engineering, STEMinist Empowered LLC Founder, Speaker, Author, Northrop Grumman Systems Engineering Manager

IG: *@drmalikagrayson*
FB: *Dr. Malika Grayson*
LI: *Malika Grayson, PhD*
Website: *malikagrayson.com*

DR. MALIKA GRAYSON is the founder of STEMinist Empowered LLC, an organization that aims to bridge the representation gap of Women of Color who pursue advanced degrees and foster a more vibrant and equitable community through collaboration, keynotes, workshops, and mentorship. To date, 22 Women have gone through the STEMinist Empowered Mentorship Program.

A global speaker and bestselling author, Grayson has given dozens of workshops and keynotes and is the recipient of many honors including SWE Advocating Women in Engineering, Zellman Warhaft Commitment to Diversity Award, Adelphi University's Top 10 Alumni Under 10 and BEYA STEM's Modern Day Technology Leader.

She also authored the best-selling book Hooded: A Black Girl's Guide to the Ph.D. Grayson's passion for increasing the number of women through the STEM pipeline motivated her to create ASPIRE STEM, which provides financial assistance to young women from high school and secondary school who aspire to pursue STEM on the university level. She also belongs to external organizations including the DiscoverE Board of Directors, the STEMNoire Planning Council, and Cornell University's Committee on Alumni Trustee Nominations.

At the time of this writing, during the day she is also a Systems Engineering Manager at Northrop Grumman!

STEM PROs Questions

Why do you do what you do? / Do you have a defining/ah-hah/ eureka moment where you knew what you wanted to do?

I have been fortunate enough to have had such an amazing STEM journey so far. I have done modeling and Systems Engineering, Software Development, R&D, IT Management and now I get to do the best of both worlds—People Leadership and Technical Leadership. I didn't have an eureka moment per say, but I knew the activities I absolutely enjoyed and really made me feel passionate like solving complex problems and helping others on their STEM Journey.

As an entrepreneur, my goal is to bridge the representation gap when it comes to women in STEM and foster a more vibrant and equitable community through keynotes, workshops, and mentorship. I am especially passionate about the advancement of women of color and providing resources, tools, and coaching.

What is something you wish you knew while in school?

I wish I knew that the major, the subjects, the career—all that I was hyper-focused on in that moment, did not define me. It was a part of me and just a portion of what would eventually contribute to who I am today. You can be more, do more, and then some. I would have reflected more on what else interested me and pursued it versus waaiting until I was done to explore more.

What is something you wish you did differently when you first started working?

I wish I was a better salesperson for myself. I had more of a keep your head down and your work will speak for yourself approach. I quickly learned that while your performance is extremely important, so is the network that you build. Those people within your network can help build your reputation for delivering results and give you exposure to opportunities.

If someone wanted to do what you do, what's the best piece of advice you'd share with them?

My day-to-day job as a People Leader and Systems Engineer/ Technical Leader depends on your passion to help others excel in their career and how you apply systems engineering concepts (e.g. requirements decomposition, verification, validation etc) to help a production program be successful.

My job as founder of STEMinist Empowered LLC is harder because you are trying to impact one person at a time. Some days it may feel like you haven't moved the needle at all, other days you are reminded that you just encouraged someone to continue their STEM path.

For either one of these roles, the best piece of advice is to remember your why. For my day to day—my why is to develop technical talent and help deliver quality products. For my company, my why is about the vision and goals I have, the women who don't see themselves represented, the young person who is now interested in pursuing STEM because they heard my story. That is my why. Everyday, assess what you learned and if you had a misstep, figure out how you could do better then try again tomorrow.

If you could definitively answer one unanswered science question, what would it be and why? / What unanswered scientific question keeps you up at night?

Is there another version of me on another planet somewhere? Is she also not a morning person? . I have always been fascinated with the unknowns of the universe and if there is another version of "our planet" somewhere. Will we one day be able to travel to it? We are learning so much already from the technology developed. I can't wait to see what more we learn.

What scientific discovery or event in history would you like to go back in time and witness?

I would just like to go back and have a conversation to understand the minds of such brilliant people. Witness their thought process from as many of our inventors as possible—understand their eureka moment.

What's the best part of doing what you do?

It's the discoveries I make daily. For example, today (on the day of this writing), I learned about radars for my job and the technology behind how they work. My background is in mechanical engineering—wind energy focus, yet I did software development, then R&D, and IT, now Systems Engineering. How fascinating it is to be able to continuously learn. You never stop.

Today, I also had a mentoring call with one of my mentees from the mentorship program I sponsor. She was encouraged to begin her dissertation writing plan after our session. That is the best part about leading an initiative I am passionate about—helping, motivating, guiding, building our future leaders.

What purchase under $100 has improved your life: career or personal?

Blackout curtains, the best neck pillow, and a daily planner (I buy one annually)

What is a necessary evil in your industry?

I want to use social media because now, it's about being known and the engagement that is deemed as necessary to help grow your business. The platforms could be such a place for positive engagement, connection and encouragement but unfortunately, it isn't always like that.

How has social media benefited or hindered your career?

I do think social media has been helpful for the growth of my brand and helping me connect to those who relate to my story and business. I have also benefited from it by being able to connect with such an awesome community of STEM professionals who are extremely supportive of each other although having never met! Although I have said social media is a necessary evil, the protected pockets of positive communities were instrumental for some challenging times.

STEM PRO

RIKHI ROY

RIKHI ROY: BSMS Aerospace Engineering, Wellness Content Creator

IG: *@rikhiroy*
TT: *@rikhi_roy*
YT: *The Leadership, Equity, and Wellness Pod*
Website: *rikhiroy.com*

RIKHI ROY received her bachelor's and master's degrees in Aerospace Engineering at the Georgia Institute of Technology, in 2020 and 2022 respectively. She is an international student from Singapore, a Brooke Owens Fellow, Aviation Week 20 Twenties Laureate, the founding Women of Aeronautics and Astronautics DEI chair under AIAA (the world's largest aerospace professional organization), and founder of Singapore's first Women leaders in Aerospace conference.

She is inspired and motivated by the concept of 'good power' which she explores through her podcast 'The Leadership, Equity, and Wellness Pod', social media platforms, and speaking engagements at organizations like NASA. She is passionate about helping students turn inward and communicate their stories meaningfully through journaling workshops across Scotland, Lebanon, Morocco, Mexico, and universities like Georgia Tech, Stanford, Yale, Columbia, MIT, and IIT.

STEM PROs Questions

Why do you do what you do? / Do you have a defining/ah-hah/ eureka moment where you knew what you wanted to do?

As a third culture kid from Singapore, my immersive educational experiences in Asia, Europe, and the United States have cultivated a passion for service and an appetite for broadening my exposure. Curious as ever, I have become a big believer in asking 'Why?' My first big 'why' was the MH370 disappearance as it revealed profound vulnerabilities in air travel in 2014. I felt a compulsion to learn more about the operation of an aircraft and use my strengths in physics and math for 'good'.

What is something you wish you knew while in school?

Sharing your journey in all your multitudes with vulnerability is a strength for deeper connection. Embracing your full self with humility and authenticity is an act of resilience for living a purpose-driven life.

What is something you wish you did differently when you first started working?

When I first started working as a graduate student in aerospace engineering, I was moved by the shift in communication styles between Research Engineers and Graduate Students. The respect, and sense of self-worth they championed and instilled was profoundly different from my formative research experiences in undergrad. I wish I had led from the belief that I was worthy regardless of the experiences I had in the past.

If someone wanted to do what you do, what's the best piece of advice you'd share with them?

Ask for help and resources to root yourself in hope, possibility, and limitlessness.

If you could definitively answer one unanswered science question, what would it be and why? / What unanswered scientific question keeps you up at night?

I was really curious about the AI impact on Aerospace Safety in graduate school and would pursue that further.

What scientific discovery or event in history would you like to go back in time and witness?

I would have loved to have been on the other side of the camera when the famous Margaret Hamilton picture was taken with all of her handwritten code.

What's a misconception about your line of work?

That there is no room for vulnerability.

What's the best part of doing what you do?

Shaping the future of human transportation, connecting people to their loved ones, safely.

What purchase under $100 has improved your life: career or personal?

Any candle I thoughtfully purchase really uplifts my spirit for anything career or personal. I love the WoodWick line that is inexpensive but has a gentle crackling sound that sounds like a fireplace.

How has social media benefited or hindered your career?

It has given me the opportunity to reach thousands of peers, students, and mentors alike who all have fascinating stories, desires and careers.

STEM PRO

DR. KENNETH HARRIS II

DR. KENNETH HARRIS II: Doctor of Engineering & MS in Program Management, BS Mechanical Engineering, Senior Project Engineer, The Aerospace Corporation, formerly NASA

IG: *@Kennyfharris*
X: *@Kennyfharris*
Website: *kennethfharris.com*

DR. KENNETH HARRIS II has been acknowledged by Forbes Magazine amongst the world's youngest and most impactful individuals in the field of science. He is known for delivering thoughtful and dynamic leadership on programs of critical national and international importance.

Kenneth is a Senior Project Engineer for The Aerospace Corporation contributing to both DoD and NASA payload missions. In this role, Dr. Harris is responsible for developing procedures and implementing processes to defend on orbit satellites from physical and cyber threats. After starting his first internship at age 16, he's worked on a total of seven missions! Including the James Webb Space Telescope (@nasawebb), the Mars Curiosity Rover, and the Artemis Program. Dr. Harris also served as a Science Show Host and correspondent for Seeker, PBS Nova and The Smithsonian capturing nearly 1M monthly views.

Kenneth graduated with honors from the University of Maryland, Baltimore County with a Bachelor of Science degree in Mechanical Engineering; he then completed a Master's degree at Johns Hopkins University in Engineering Management and a Doctor of Engineering degree at George Washington University.

STEM PROs Questions

Why do you do what you do? / Do you have a defining/ah-hah/ eureka moment where you knew what you wanted to do?

A huge part of my journey is my Father, who is a mechanical engineer. I was fortunate enough to have a role model like him around all the time. Now, I wasn't forced to be an engineer by him, or even strongly suggested. But, having a support system like a dad that walked the same path as you is amazing. I remember my ah-hah moment vividly. The first time I was bit by the "space bug" was in 1998 I was 6 years old leaving the office one evening with my dad, and hanging on the wall was an image of The Pillars of Creation. My curiosity swelled and every day since then I've wondered "How small are we in this vast universe?"

What is something you wish you knew while in school?

If there's one thing I wish I knew while in school, it's the importance of living in the moment and cherishing each experience. When you're young, it's easy to get caught up in the hustle and bustle of academic pressures, extracurricular activities, and the relentless march towards the future. But looking back, I realize that every moment, every class, and every interaction played a crucial role in shaping who I am today. Rather than constantly worrying about what's next, I wish I had taken the time to fully immerse myself in the present, to savor the joys and challenges of each grade, and to appreciate the journey of learning and growth. Every experience, whether it was a triumph or a setback, served as a valuable building block in my personal and academic development. By embracing the journey and living in the moment, I could have fully embraced the richness of my educational experience and cultivated a deeper sense of gratitude and fulfillment along the way.

What is something you wish you did differently when you first started working?

When I first started working, I wish I had prioritized seeking feedback more actively. Feedback is incredibly valuable for growth and improvement, and early in my career, I didn't always seek it out as much as I should have. Embracing feedback earlier would have helped me identify areas for improvement sooner and allowed me to develop professionally more quickly. Also mentorship has always been big to me, but because I was so reserved I didn't pursue it as frequently as I should have.

If someone wanted to do what you do, what's the best piece of advice you'd share with them?

If you're eyeing a career like mine, my best advice? Find yourself some accountability buddies ASAP. This field can throw some serious curveballs, and without a solid support crew, you might be tempted to throw in the towel. Having someone or a group to cheer you on, pick you up when you're down, and keep you on track is priceless. And let's talk about failure. It's not just a possibility; it's practically a rite of passage. But here's the thing: how you handle it and who you lean on makes all the difference. When you've got your squad backing you up, setbacks become more like speed bumps than roadblocks. So, don't go it alone. Build your tribe, embrace the bumps, and keep pushing forward. You've got this!

If you could definitively answer one unanswered science question, what would it be and why? / What unanswered scientific question keeps you up at night?

How unique are we in the universe? When you really think about the complexities of everything that exist on this planet from humans to plants, to even how the weather functions could this exist on another planet. Do they have the same limitations as us and is that why we haven't encountered them or vice versa. The question of how unique life is in the universe intrigues me deeply. I often wonder if the complexities of life on Earth, from humans to ecosystems, exist elsewhere. Are there other planetary systems with similar conditions and limitations, or have life forms evolved differently? Despite our technological advancements, the vastness of space means we've yet to find answers. But exploring this question could revolutionize our understanding of the cosmos and our place within it.

What scientific discovery or event in history would you like to go back in time and witness?

One scientific discovery I'd love to witness firsthand is the unveiling of the double helix structure of DNA by James Watson and Francis Crick in 1953. This momentous event not only revolutionized our understanding of genetics but also laid the foundation for countless breakthroughs in medicine, biology, and biotechnology. Witnessing the excitement and realization of the implications of this discovery would be truly awe-inspiring.

What's a misconception about your line of work?

A lot of people think we're constantly building hardware or preparing for a launch. But, there's years of planning that go into it before we actually throw on our bunny suits and get into the cleanroom. The general public also assumes we're hiding information about aliens and special technology!

What's the best part of doing what you do?

One of the best parts of my job isn't even in my job description. Mentorship and engaging with the next generation is a truly rewarding experience. It's like being a trusted guide on a professional journey, offering support, wisdom, and encouragement along the way. When I mentor, I take pride in sharing my knowledge and experiences to help others navigate their career paths. It's fulfilling to see my mentees grow and develop, gaining confidence and overcoming challenges with my guidance. Moreover, mentorship is a mutual exchange. While I offer advice and insights, I also learn from my mentees' unique perspectives and experiences. It's a dynamic relationship that enriches both parties and fosters continuous growth and development. Ultimately, mentoring is about building meaningful connections and investing in each other's success. It's a privilege to play a role in shaping the next generation of professionals and contributing to a supportive and collaborative community.

What purchase under $100 has improved your life: career or personal?

During my very first internship, my mentor offered me a simple yet invaluable piece of advice: invest in a journal to log my daily activities. At first, it seemed like a small and mundane suggestion, but little did I know how profoundly that $10.00 piece of advice would impact both my career and personal life. With my new journal in hand, I began diligently recording my tasks, accomplishments, challenges, and insights each day. This practice not only helped me stay organized and focused but also provided a tangible record of my professional growth and accomplishments over time. From a career perspective, keeping a journal allowed me to track my progress on various projects and assignments, identify patterns in my work habits, and reflect on my strengths and areas for improvement. It helped me set goals, prioritize tasks, and hold myself accountable for my actions. Moreover, having a documented record of my achievements proved invaluable when it came time to update my resume or prepare for performance reviews and job interviews.

How has social media benefited or hindered your career?

Social media has been a powerful tool for me in representing diverse groups in engineering. Through platforms like Twitter, LinkedIn, and Instagram, I've been able to share my experiences, insights, and achievements as a member of underrepresented communities in the field. By amplifying my voice and showcasing my work, I've had the opportunity to inspire and empower others who may not see themselves reflected in traditional engineering spaces. Additionally, social media has connected me with like-minded individuals and organizations committed to diversity and inclusion, enabling me to collaborate on initiatives and projects that promote equity and representation in STEM fields. Moreover, social media has allowed me to leverage my platform to advocate for systemic change within the engineering industry. By raising awareness of issues such as gender bias, racial discrimination, and accessibility barriers, I've sparked important conversations and pushed for greater accountability and inclusivity in engineering spaces. Through sharing resources, advocating for policy reforms, and amplifying the voices of marginalized communities, social media has enabled me to drive meaningful change and contribute to a more equitable and welcoming environment for all engineers, regardless of background or identity.

STEM PRO

DIANA APONTE

DIANA APONTE: Colombian Musical Artist bridging Art & Science

IG: *@dianaaponte*
X: *@dianaaponte*
FB: *Diana Aponte*
YT: *@dianaapontemusic*

DIANA APONTE is a Colombian artist renowned for her ability to bridge the worlds of art and science. Her approach to merging these disciplines has been reflected in various projects in the aerospace, musical, and social industries. She has collaborated with the United Nations, NASA Artemis Moon Snap, Conex Research, International Astronautical Federation, BOS Planet, Sidelaris Foundation & The Inspired 24.

Diana creates music inspired by science and space and is currently working on her new album in collaboration with scientists around the world. Diana has performed at international space congresses and is an active ambassador promoting the diversification of the space industry.

Beyond her art, Diana is diligently involved in social projects promoting STEAM education throughout Latinoamérica and Europe. Her work inspires people of all ages to explore new possibilities and look towards the stars through art.

STEM PROs Questions

Why do you do what you do? Do you have a defining/ah-hah/ eureka moment where you knew what you wanted to do?

When you grow up being inspired by the night sky and music, it is inevitable to force yourself to find a logical explanation of why those passions are there. You want to explain the purpose of our place in the universe as much as you want to find your truest self through art. In most of our entire lives we've been taught that we have to choose just one way or the other, but me and other artists around the world are rebel enough to fight against the odds to create new ways to explore this amalgam full of uncertainty and excitement. My mission is to bring this vision and projects to Colombian and Latin-American young people so they can envision themselves exploring space and art because that is one of the things the world needs the most right now. Passion to communicate what makes us humans and explore new ways to unite us though art for a better world is my why.

I had a very hard health crisis years ago, and I was thinking if I was about to die, what would be my biggest regret in this life. I reflected on this question for days, even weeks, until I checked the SpaceX video of Elon Musk and Yusaku Maezawa calling for artists to apply for the Dear Moon Project, and it came to me as a clear answer. There, I knew that this path is the one that I had to follow deep in my heart and that it was possible to pursue even though it was against my previous plans and the conventional ways. That I found the light but now it was my job to build the way to get there.

What is something you wish you knew while in school?

That not fitting in didn't mean that something was wrong or that I was a bad person. Not being so hard on myself, understand that failure is not fatal and is completely necessary to keep growing and learning. It's fun to fail sometimes. The best you can do is fail as much as you can in as many things as you can. That will lead you to the right place at the right time.

What is something you wish you did differently when you first started working?

Trusting more in my instincts and ideas. The confidence it takes to develop your art of any project is something you need in order to honor your vision and its purpose. If it sells or fills others expectations is not your problem. Focus on the work, do it with discipline and it will find its own way to the right places and the right hearts in its own timing. Hurts in the ego but this concept also brings peace to create freely without attachments to the outcome.

If someone wanted to do what you do, what's the best piece of advice you'd share with them?

Take the first step now, don't wait, just create! Don't be afraid to be the first doing the things you ambition, even if it's risky, Please do it, do it with what you can, how you can, at your own phase, don't limit yourself according to the things that others have made before you, follow the most exciting path and every day take one little step towards the things that excites you and makes you feel alive. Enjoy all the simple things along the journey, when you breathe, your coffee, the sip of water, your face in the window feeling the air, your pet watching you from the other side of the room. Do it, and create your art on your own terms, if you have to rest, do it. Resting, having space and silence is crucial in any creative process and if you trust and put your 100% of focus, your love and authenticity, the wind will go in your favor and your art will find its own place.

If you could definitively answer one unanswered science question, what would it be and why? / What unanswered scientific question keeps you up at night?

The unified field theory. Who doesn't think about it every night? It's as exciting as terrifying how little we know about the place we are living in. Sometimes my head goes to the space of how to help to resolve this question. I am sure that is not a one person's job, but a project for an entire team. I'd like to get involved but don't really know how you can even start. I will get more involved in mathematics in the incoming years because I feel it's very similar to music in many ways I love and I see it as a beautiful language I would like to explore to use and compose. Definitely on my bucket list, not to resolve this unanswered question but at least to try to approach the problem from a different and poetic perspective.

What scientific discovery or event in history would you like to go back in time and witness?

If I were to choose a scientific discovery or event in history to witness firsthand, it would undoubtedly be the day the Higgs boson was discovered. CERN is one of my favorite places on earth and it holds a special place in my heart. The atmosphere is filled with excitement and possibility, and the people I've encountered there are not only brilliant but the best human beings I have ever met. There's an indescribable energy at CERN that transcends words, and just imagining the moment scientists unveiled the existence of the Higgs boson sends shivers down my spine. Everyone should go at least once to CERN to experience it.

What's a misconception about your line of work?

In Colombia and Latin America, that it doesn't exist. We haven't had any remarkable artists working in the space sector that we can have as examples or leaders so most of the people think it is a line of work that does not exist. Around the world I've had amazing experiences meeting artists that are working on the space sector and combine beautifully art and science. We inspire each other and I will make sure that the new generation can have this guiding light on their journey so they know it's not only possible but required in our world and that we as Latin American people also belong to this industry.

What's the best part of doing what you do?

The best part of mixing art and science is the possibility to think outside of the box in the creative process and the people I get to work with. I have never felt more at home as with my space colleagues, we have different backgrounds and nationalities but the same passion, desires, and life experiences. For me collaboration is the most beautiful part of this journey.

What is a necessary evil in your industry?

Diplomacy. To convey so many points of view, different cultural backgrounds, religions and political preferences, diplomacy is crucial to keep focusing on the thing that matters, our common direction, and reminding what unites us as humans .

Have you ever changed your line of work? If so, why and what was the change?

Yes, many times and I think it is healthy every couple of years to rethink your purpose and the way you want to achieve it, to adjust accordingly if necessary. I was a musician in a symphonic orchestra, then I became a jazz singer in a big band, then a song writer doing social work and now I am working as an artist in the space industry and continuing with social work which is part of who I am. To whoever is reading this book, don't be afraid to follow the desire of your heart, it contains the key to your real happiness and the pathway to your purpose. Do it for you, and do it for us.

What purchase under $100 has improved your life: career or personal?

An ukulele I bought for 50USD to my daughter, she learned by herself and she has been playing everyday day 1 has been writing a lot of songs with it, she loved it. I recommend everyone to purchase any musical instrument. It makes your life more beautiful. Also using potassium-magnesium citrate everyday has changed my life for the better, making my heart stronger.

How has social media benefited or hindered your career?

It's been crucial to connect with leaders from other countries, collaborate and receive opportunities in the space industry. I love the sense of community we have and how we support each other. When you curate your social media as an art gallery, you will feel inspired everyday with the work that others are doing and sharing. I'd love to recommend 3 of my favorite scientific artists on social media, Dr. Merrit Moore @merrittmoore the quantum ballerina, Cathrin Machin @cathrinmachin space painter and nebula artist, and Reuben Margolin @reubenmargolin who makes kinetic sculptures inspired by nature and mathematics. Social media is the most important platform to develop your career and show the world who you are and the mission of your art and work. Don't be scared by numbers, trends and algorithms, instead be inspired by human connection and a world of possibilities. So please don't hide yourself, we want to find you and support you, there is enough space in STEAM for all. If you truly believe that the sky is not the limit you are already one of us. Welcome to the club!

STEM PRO
DR. ALLIE (FOLCIK) GOBEIL

DR. ALLIE (FOLCIK) GOBEIL: Environmental Toxicologist, Science Communicator, Fashion Blogger

IG: *@allie.gobeil*

DR. ALLIE (FOLCIK) GOBEIL is a toxicologist working in environmental consulting with a passion for science communication.

As a kid, she always had a love for the sciences and originally dreamed of exploring the stars as an astronomer. But as she got older, her interests skewed to the terrestrial, developing a passion for environmental science and sustainability.

Allie completed her undergraduate studies at Florida Institute of Technology in Biochemistry with a minor in Sustainability. Being at a smaller school, she had the opportunity to start working in various labs developing her research skills. After experimenting with Crustacean Ecology (mainly taking care of horseshoe and stone crabs) and Paleoecology[1] (counting

1 *https://www.scielo.br/j/bn/a/cGQyjDsPryfVByv7RFwm6Dc/?lang=en*

fungi in sediment cores to estimate megafaunal extinction), she found her fit in Dr. Palmer's Plant Biochemistry lab working with plants and algae. Her undergraduate project focused on developing methodology for small particle tracking for use in investigating quorum sensing behavior in unicellular algae.[2] Entering research early in school led to Allie being named an Astronaut Scholar in 2016.

Allie happened upon the field of toxicology while attending an internship with the National Center for Toxicological Research (NCTR) under the FDA during her Junior year. By presenting her research at the Society of Toxicology (SOT) annual meeting, she was connected with professors at Texas A&M which she then attended for her Ph.D. in Toxicology. Allie's doctoral research focused on using a form of ionizing radiation called electron beam (eBeam) technology to clean up water that is contaminated with toxins produced by harmful algae and bacteria (see the algae overlap?).[3]

In 2020, Allie started a side project called The Pretty PhD Blog initially fueled by comments received through the years about being a female in science. Her goal was to help break scientist stereotypes and show aspiring female scientists that you can be educated while also having a social life and being fashionable (another side hobby).

As of 2024, Allie is a Managing Scientist at Exponent working as a toxicologist across multiple disciplines including product stewardship, risk assessment, natural resource damage assessment, and regulatory compliance.

STEM PROs Questions

Why do you do what you do? / Do you have a defining/ah-hah/ eureka moment where you knew what you wanted to do?

I wouldn't say I had an ah-hah moment leading me to what I do now. Rather, I think it was the accumulation of all of my experiences that built my foundation. I'm someone who has always felt strongly about applying science and things that I've learned to real world situations. Of course we

2 *https://www.cell.com/iscience/pdf/S2589-0042(20)30911-1.pdf*

3 *https://oaktrust.library.tamu.edu/bitstream/handle/1969.1/196279/FOLCIK-DISSERTATION-2021.pdf?sequence=1&isAllowed=y*

need scientists that want to work in a lab and figure out the fundamentals of why things happen, but what I like most about what I do is that I'm able to help somebody solve a problem they have. A lot of times it requires thinking out-of-the-box and bringing in all different types of sciences.

What is something you wish you knew while in school?

So many things! But the thing at the top of my list is that I wish I knew I didn't need to have my career decided at the age of 18! I definitely had some direction—knowing I wanted to be in science—but it took me time and experimenting to narrow down what I wanted to do. I switched majors, worked in different laboratories, and changed my mind a lot, but that eventually led me to a career I love (and a very multidisciplinary background).

What is something you wish you did differently when you first started working?

I would say I wish I had more confidence to speak up and share my ideas—something I still struggle with. Imposter syndrome never goes away! But I wholeheartedly believe that a team is made stronger by the different experiences and diversity of the people that are on it. Sharing your thoughts and opinions not only helps you grow in your career but is also important for the team.

If someone wanted to do what you do, what's the best piece of advice you'd share with them?

Consulting is a field that isn't really talked about in school. I was initially introduced to consulting through my graduate advisor who had a project in conjunction with some researchers at an environmental consulting firm. What drew me to the field was the idea that you're always working on different projects and never doing the same thing. It's very much an application of everything that I've learned. For someone who is interested in going into consulting, my best piece of advice is to make sure you're strong on your core sciences like biology and chemistry. You're expected to become an expert at one thing in grad school, but having the flexibility to learn something new will allow you to take on so many new and interesting projects. In terms of actually getting into the field itself, networking is super important. Make sure to introduce yourself to people/companies at conferences and events to get your name out there, even if you're not actively looking for a job.

If you could definitively answer one unanswered science question, what would it be and why? / What unanswered scientific question keeps you up at night?

How to make French fries less caloric? Just kidding. Maybe not a single question but I wish I knew how to better predict the toxicity of something. Often we don't know that a chemical or some other type of environmental exposure is toxic until way farther down the road when we start seeing an effect. I wish we could better characterize this toxicity in order to prevent cancers, environmental pollution, etc.

What scientific discovery or event in history would you like to go back in time and witness?

I think it would be pretty crazy to be in the mission control room when Apollo 11 landed on the moon!

What's a misconception about your line of work?

A huge misconception in the field of toxicology is the idea that being exposed to something means you're at risk. The doctrine of toxicology is, "the dose makes the poison", meaning that factors other than just the presence of a substance contribute to its potential risk. For example, we don't think of water as toxic, but too much of it can be. Obviously it's more complicated than just the amount, but the principle remains the same.

What's the best part of doing what you do?

The best part of doing what I do is really just being able to continuously learn new and interesting things.

What is a necessary evil in your industry?

A necessary evil in my industry is that lots of work ends up being reactive, meaning in response to something that happened. This might be in response to a lawsuit, a regulation, or a natural disaster. It's always great to work on proactive projects though when a client is trying to make something better or get ahead!

Have you ever changed your line of work? If so, why and what was the change?

I am still new in my career so I haven't made any line-of-work changes yet. As much as I love the flexibility and variability of the work that I do, it is easy to overwork yourself. So, in the future if I did make any changes it would likely be to balance out my work life balance a little bit!

What purchase under $100 has improved your life: career or personal?

A super inexpensive purchase that I love was an electric mug warmer on Amazon. I keep my coffee on it on my desk at work so it doesn't go cold when I forget about it. Caffeination is important!

How has social media benefited or hindered your career?

Social media has been an amazing tool that has allowed me to make so many connections with other scientists. I was first exposed to science communication in undergrad while working on the inaugural Youth Making Ripples Film Competition which taught and encouraged ocean conservation to K-12 students. Social media wasn't really big yet, but the idea of combining creative aspects like film with science was really interesting to me.

Through blogging in graduate school, I was able to create a community and support system that really helped me get through some of the harder times of my Ph.D. It did become more overwhelming to produce content for my blog once I left school, but reflecting on all the effort I put into my platform, I am thankful for the friends made and also the encouragement from strangers thanking me for sharing my journey. Without my social media community, it would have been much harder to finish school and get to my current career.

Social media and groups like Women in Science and Engineering (WISE) also helped connect me with groups of students where I was able to share my background and encourage STEM as a career path. Developing presentations and blog content really helped hone my communication and writing skills and has been very beneficial to my career in consulting.

I hope I was able to convince at least one person that the struggles of grad school or science in general are worth it!

STEM PRO

KENNEN HUTCHINSON

KENNEN HUTCHINSON: Science Communication and Information Design Consultant

IG: *@kennenhutchison, @withkennen*
YT: *@sciencewithkennen,*
@acquaintedworldmedia
LI: *@kennenhutchison*
SC: *@kennenhutchison*
Websites: *acquaintedworld.com,*
withkennen.com (coming soon)

KENNEN HUTCHISON is the Founder and CEO of Acquainted World, LLC (www.acquaintedworld.com), a company dedicated to public science communication and information design, where he works as a research and communication consultant. He's also a lecturer of *Disease Dynamics* at the School of the Art Institute of Chicago, the creator and host of @withkennen, and one of my wonderful former cohosts from @thekingofrandom. Kennen has a Bachelor of Science in Microbiology from Western Illinois University, a Master of Arts in Learning Sciences from Northwestern University, and 10 years of research experience.

STEM PROs Questions

Why do you do what you do? / Do you have a defining/ah-hah/ eureka moment where you knew what you wanted to do?

Absolutely! My journey into the world of science communication is fueled by a deep-seated desire to make science relatable and exhilarating for everyone. My pivotal moment? It was split into two unforgettable instances. The first spark was when I took my first microbiology class and something just clicked that opened my eyes to how all of life is connected. I suddenly realized how understanding science helped me not only to understand myself but the whole world around me. The second was when I managed to simplify my complex research for a friend. The look of excitement on their face as the concepts clicked and the knowledge fell into place was just priceless. That was it for me; I knew this was what I wanted to do—help others hit those "ah-ha" moments as they discover the beauty and practical importance of the world around them!

What is something you wish you knew while in school?

Oh, if I could have a chat with my younger self, I'd say, "Guess what? You don't need a PhD to make waves in STEM." It's a lesson that really would have reshaped my early journey and broadened my view on the countless paths one can take in the fascinating world of science. It also would have saved me a lot of stress and depression.

What is something you wish you did differently when you first started working?

Back in 2017, I took a detour from my dream of launching a science communication show to pursue a Ph.D. that didn't fully light a fire in my heart. Looking back, I'd tell myself to stick to what truly sparks joy and excitement within me—creating content that bridges science and society.

If someone wanted to do what you do, what's the best piece of advice you'd share with them?

If someone wanted to do what I do, I'd give them this advice: Stay wildly curious about the world and the people in it—never lose your excitement to learn new things. Be humble and embrace the fact that you could be wrong about everything—it's okay to be wrong, but it's not okay to not grow from new knowledge. Practice your storytelling skills because that's how you connect. And lastly, learn to analyze and criticize data—learn how knowledge is created and what makes research rigorous. These pillars, supported by a foundation of solid scientific understanding, will be your guide.

If you could definitively answer one unanswered science question, what would it be and why?

The origin of life keeps me up at night—it's the ultimate puzzle. With so many hypotheses floating around, wouldn't it be something to uncover the real story? It's a question whose answer could transform our understanding of life itself.

What scientific discovery or event in history would you like to go back in time and witness?

Time travel isn't on my wish list, but I'd love a safe, behind-the-scenes peek at how humanity's early architectural wonders, like the pyramids or Stonehenge, were constructed. The ingenuity and determination must have been truly awe-inspiring.

What's a misconception about your line of work?

Some think that scientists, and by extension science communicators, are just about the facts and figures, missing the human touch. On the contrary, our drive to research and share scientific knowledge is deeply rooted in our passion for helping others and attempting to improve the human experience.

What's the best part of doing what you do?

There's nothing quite like the moment someone's eyes light up with understanding—when they see the world a little differently that can make their lives a little better because of something we've shared. It's those moments of clarity, connection, and impact that make everything worth it.

What is a necessary evil in your industry?

Definitely, funding. It's the fuel for scientific explorations but also a source of brutal competition which can lead to toxic work environments and dubious decision-making. Balancing the pursuit of knowledge with the reality of limited resources is a constant challenge that always has to be considered.

Have you ever changed your line of work? If so, why and what was the change?

Only about 100 times! I've worn many hats over the years—teacher, coach, content creator, researcher, consultant, entrepreneur,even construction worker—you name it! I enjoy learning through doing, so when I'm interested in a career, rather than just reading about it, I try to go out and experience it firsthand. Each role taught me something invaluable about myself and the world, guiding me to where I am today. This diverse experience has not only enriched my understanding but also given me a unique toolkit for engaging with a wide audience.

I started on a medical degree track but left to work in construction when I realized I didn't want the daily pressure of a wrong call costing a patient their life. While I loved construction, I missed studying science, so I went back for a degree in science education. While working as a teaching aide, I realized I didn't like the constraints of K-12 education, but I did love learning about microbiology, so I got my degree in microbiology. Not sure what to do with a degree in microbiology other than research or teaching at a college, I went on to pursue a PhD in medical microbiology which I thought was needed to do both. I learned that I neither enjoyed the work nor the environment of my biomedical PhD, so I changed paths to an educational degree to get a Master's instead. Collectively, these experiences taught me that I love teaching science (STEM) to people who can directly and practically use that knowledge to improve their lives. Furthermore, they taught me the importance of working for myself or from a position of control so that others cannot compromise the rigor or ethics of my work. This led to starting my own science communication media company, which then became a science communication consulting company.

What purchase under $100 has improved your life: career or personal?

In today's world, finding anything under $100 seems like a challenge. But the most transformative investment I've made has been in my mental health. Regular therapy and mindfulness practices have been game-changers in my personal and professional growth, bringing some calmness and clarity amidst life's chaos that far exceeds any monetary value.

How has social media benefited or hindered your career?

Social media is a double-edged sword. On one hand, it's an incredible platform that has been instrumental in building connections and sharing my work—for connecting with people from all corners of the globe and sparking meaningful discussions. On the other hand, it has served as a breeding ground for imposter syndrome, providing an artificial source of stress, with the pressure to present a polished image of ever-elusive perfection. Striking the right balance is key. Balancing its benefits with its emotional toll remains an ongoing challenge for me even today.

STEM PRO
SARAH ROBERTS

SARAH ROBERTS: Environmental Journalist, Author and Biologist specialising in the animal behaviour of big predators.

IG: *@sarahsrealjob*
TT: *@sarahsrealjob*
X: *@sarahvroberts*
FB: *Sarahsrealjob*
YT: *@sarahsrealjob*
Website: *sarahrobertsofficial.com, thisiscreature.com*

SARAH ROBERTS is an Environmental Journalist and Author, who begun her career in Animal Behaviour focusing on big predators. She originally worked out in Shark Lab before being bitten by a small lemon shark and writing her first children's book, Somebody Swallowed Stanley (all about a plastic bag confused as a jellyfish which was recently read out by Tom Hardy on CBeebies bedtime stories).

Sarah's job has taken her to some of the most remote places in the world, working as a grizzly bear guide on a floating lodge, to filming a doc about climate change technology in Iceland. She is passionate about protecting the planet and uses storytelling to do this. She created an outreach project, Creature, in 2014 and spends a portion of each year visiting schools around the world ever since.

Sarah's journalism includes pieces on shark fishing, human wildlife conflict with tigers and more recently a 6 part podcast series on the future of the British uplands.

STEM PROs Questions

Why do you do what you do? / Do you have a defining/ah-hah/ eureka moment where you knew what you wanted to do?

I've always loved animals, I've always loved adventures and I've always loved stories. Originally I had the idea that I could just live off grid with my surfboard in a beach hut with my husky and work with wildlife for the rest of my life. I started off working in field research around the world, but no matter how far I traveled, I always found human impacts. I quickly realised that my favourite places and favourite wildlife species are not going to be there for the rest of my life, if we don't all do something to change our global behaviour. So that's when I moved into journalism and book writing.

What is something you wish you knew while in school?

It's not what you know, it's who you know.

What is something you wish you did differently when you first started working?

I don't think I would change anything from the past. Although it's been a winding road, every knock back or closed door has got me to where I am now. I think mistakes are there to be learnt from, and sometimes when you think you're failing, it's actually another door that's opening for you.

If someone wanted to do what you do, what's the best piece of advice you'd share with them?

Don't be put off by the amount of people that tell you that you can't do it. The reason that my channels are called 'Sarahsrealjob' is because I spent a lot of my career being told that 'one day I would need to get one'. If you have a vision of how you want your life to be, be bold and stick to it, see it through.

If you could definitively answer one unanswered science question, what would it be and why? / What unanswered scientific question keeps you up at night?

There's so many big questions that are already being investigated and I am eagerly waiting to see the results, such as: will we successfully reverse aging? or will AI help or destroy the human race? Probably the main question that I often find myself thinking is 'what is outside of space?'

What scientific discovery or event in history would you like to go back in time and witness?

There's so many to choose from, but as somebody who has spent their career focused on big predators, it's got to be at the dinosaurs!

What's a misconception about your line of work?

I don't think people realise it took over 5 years of knock backs before I could get my most successful children's book accepted by a publisher. Since then it's been printed around the world in multiple languages and has been read by Tom Hardy on tv. But from the outside, it looks like an overnight success.

What's the best part of doing what you do?

I get to meet amazing people all over the world and hear their stories and firsthand animal encounters. I get access to places that tourists can't go. And I get to see wildlife in some of the most natural, exotic and extreme environments.

What is a necessary evil in your industry?

Mosquitoes, sleepless nights, jetlag, lots of working for free and suspect looking food (sometimes).

Have you ever changed your line of work? If so, why and what was the change?

I started in field research working with sharks. After I got bitten, I couldn't go in the water and had to find another way of being useful. That's when I started writing my children's book and started my education outreach company as well. When I worked in the African bush, my job was in field research and I was alone all day, so I started filming to keep myself sane. That's when I started my YouTube channel for the first time. By the time I went to Canada to work as a grizzly bear guide, I realised that storytelling and journalism was where my heart was at. Since then I've split my time documenting different environmental issues in the field and taking what I have found into schools all over the world through books and workshops.

What purchase under $100 has improved your life: career or personal?

A notebook and pen

How has social media benefited or hindered your career?

It has given me a platform, and the freedom to tell stories about the things I care about without restrictions. It's the reason I am able to do a lot of my career. But social media is also incredibly time-consuming, can cause a lot of stress and because there are so many people with their own platforms, it dilutes the reach of many of these stories and makes it harder to find a budget at times.

Additionally, a lot of the wonderful and amazing places around the world are under threat because these locations are being shared online and people are traveling to them and unwittingly disturbing the local habitats and wildlife. It's definitely a double-edged sword.

"An investment in knowledge pays the best interest."

—Benjamin Franklin, American Founding Father, scientist, inventor, diplomat, statesman, and writer known for numerous inventions, including the lightning rod

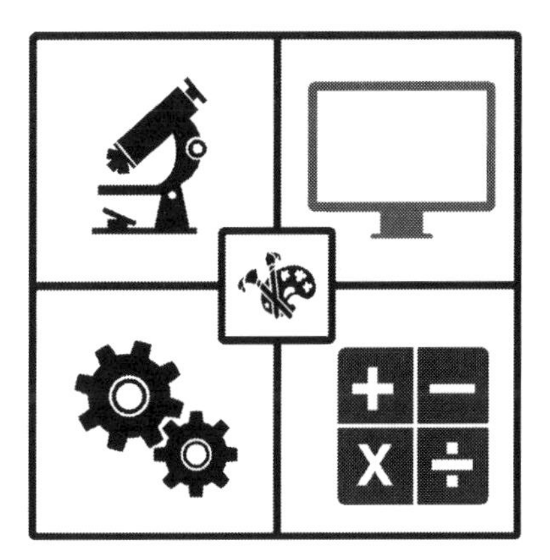

STEM PRO
DR. SAMANTHA BALLARD

DR. SAMANTHA BALLARD: Scientist/Engineer, PhD Atmospheric and Oceanic Physics, BS Atmospheric Sciences, PostDoc with CSTARS and the SUrge-STructure-Atmosphere INTeraction (SUSTAIN) Laboratory

IG: *@dr.air.sea.sam*
TT: *@dr.air.sea.sam*
X: *@air_sea_sam*
YT: *@samanthaballard6038*

DR. SAMANTHA BALLARD is a multi-disciplinary scientist/ engineer. She obtained her PhD in atmospheric and oceanic physics (remote sensing/satellite speciality) from the University of Miami in conjunction with the Center for Southeastern Advanced Tropical Remote Sensing (CSTARS). She received her B.S. in atmospheric sciences from Penn State University. She has worked as a postdoctoral research associate in partnership with CSTARS and the SUrge-STructure-Atmosphere INTeraction (SUSTAIN) laboratory.

Samantha's PhD research focused on a pilot experiment called the Coastal Land Air Sea Interaction (CLASI) experiment which utilizes satellite imagery and field instruments to improve atmospheric and oceanic modeling. CLASI was funded for a multi year follow on experiment for longer term study of coastal environments in California and Florida. She founded air sea analytics LLC. Samantha's day to day job involves working on government contracts. experiment planning, executing, programming, writing, system engineering, satellite and field instrumentation data analysis. She has also held internships with NASA working on developing satellite algorithms.

STEM PROs Questions

Why do you do what you do? / Do you have a defining/ah-hah/ eureka moment where you knew what you wanted to do?

I do what I do because I love learning about how the earth works around me and I want to share that knowledge with and inspire others. I've been interested in science, weather and the ocean particularly since I can first remember. My interest started as a fear of storms and the ocean growing up. I realized that in order to conquer those fears I needed to learn more about the weather and the ocean and those fears turned to love!

What is something you wish you knew while in school?

Nothing is as scary as you think it is. Also communication skills are just as important as technical skills and they will get more important with every step you take.

What is something you wish you did differently when you first started working?

Documenting. I started to compile my successes and failures which I started doing on social media which wasn't as prevalent when I first started working.

If someone wanted to do what you do, what's the best piece of advice you'd share with them?

Be curious about your surroundings and how things work in your daily life. Tinker and think out in nature and with tech/machines in your daily life. And go for a swim/walk on the beach.

If you could definitively answer one unanswered science question, what would it be and why? / What unanswered scientific question keeps you up at night?

Improved coastal weather/ocean forecasting and satellite observations for climate. Also what science we can study from our sailboat.

What scientific discovery or event in history would you like to go back in time and witness?

First weather satellite launch, TIROS I in 1960.

What's a misconception about your line of work?

That I'm a marine biologist. You can work with them though and collaborate on projects! And engineers, software engineers, and other physicists.

What's the best part of doing what you do?

Learning new things everyday and being out in the field with instrumentations or on the ocean!

What is a necessary evil in your industry?

Programming. Not always evil but sometimes.

Have you ever changed your line of work? If so, why and what was the change?

More robotics

What purchase under $100 has improved your life: career or personal?

Paid weather apps

How has social media benefited or hindered your career?

Social media has been an excellent networking tool that has benefited my career and given me many opportunities work wise. It has given me amazing friends and colleagues as well!

STEM PRO
ASH WHEELER

ASH WHEELER: Astronomical Artist, Dust & Ash Co., Recent Astrophysics Graduate

IG: *@dustandashco*
TT: *@dustandashco*
X: *@dustandashco*
FB: *Dust & Ash Co.*

ASH WHEELER is an astronomical artist living and working in Atlanta, Georgia.

She is currently painting the spherical bodies of our solar system in conjunction with completing her degree in astrophysics (as of May 2024) in order to support her art and work in the space industry.

STEM PROs Questions

Why do you do what you do? / Do you have a defining/ah-hah/ eureka moment where you knew what you wanted to do?

I do what I do simply because I love space. Astronomy has fascinated me from an early age and I have found no higher calling in my career than exploring the cosmos (and sharing what I know)! I realized at the tail-end of my art degree that what I really wanted to pursue in life was astronomy/astrophysics. My love for the moon motivated me to make a drastic academic and life change—I don't regret it one bit.

What is something you wish you knew while in school?

Something I wish I knew earlier on in my physics degree was the importance of the academic community. I struggled greatly in my early calculus classes and felt very alone in that—little did I know I was not the only one having a hard time! However, I didn't find that out until I became involved with STEM groups on campus and started talking with my peers. Having a group of people to share frustrations with and brainstorm solutions was the greatest thing I did in order to thrive in my upper-level courses.

What is something you wish you did differently when you first started working?

I don't know if I have an answer for this since my business was very niche at its conception and I worked as hard as I could in order to get it off the ground. Maybe with time and experience I will come to realize things I could have changed!

If someone wanted to do what you do, what's the best piece of advice you'd share with them?

Plug into the STEM community online and be inspired by what other people have accomplished and how *unique* all of their journeys are! Don't compare yourself to others, there is no linear path to being "successful" in art and astrophysics.

If you could definitively answer one unanswered science question, what would it be and why? / What unanswered scientific question keeps you up at night?

"Is there life beyond earth?" would be a science question I would love to answer one day. I believe that singular question is something most people have asked themselves at some point in their life and to be able to confirm (because I believe there is!) this would bring such great purpose and sense of satisfaction in my own life. It would also help me sleep better at night!

What scientific discovery or event in history would you like to go back in time and witness?

Without a doubt, the Apollo 11 Moon Landing!

What's a misconception about your line of work?

Regarding art and science, a misconception I personally had is that I couldn't be "good" at both. I grew up with a natural inclination/talent for art and I struggled greatly with STEM subjects. I never thought I could succeed in a scientific career because of this, but I proved myself wrong when I put forth a greater effort and surrounded myself with people that supported me in that endeavor.

What's the best part of doing what you do?

The best part of creating art that mimics the beauty of nature is getting to study it as I go—it never ceases to amaze me how planets and moons evolve and thrive in environments so foreign in comparison to our own! The best part is that it is so available to us via modern technology, waiting to be explored and admired.

What is a necessary evil in your industry?

In art and in school, failing is a tool. You can only excel to great heights when you know what it is to be wrong and what you need to change.

Have you ever changed your line of work? If so, why and what was the change?

I'm not sure if I need to answer this since the main artery of my business/school journey is this very question :)

What purchase under $100 has improved your life: career or personal?

I'm not sure if I have an answer for this beyond.... a lot of coffee.

How has social media benefited or hindered your career?

Social media has benefited my career in such a way that my business wouldn't exist without it! I have made so many friends on the internet through my love for space and that has connected me with an audience who appreciates the merging of space and art. People all over the world have supported my business because of my social media platforms and for that, I am beyond grateful.

STEM PRO

DR. JANINA JEFF

DR. JANINA JEFF: PhD Human Genetics, Staff Bioinformatics Scientist at Illumina, Podcast Host "In Those Genes" @inthosegenespod

IG: *@djaysquared*

DR. JANINA M. JEFF is a population geneticist, bioinformatician, STEAM-activist, educator, motivational speaker, and podcaster, Dr. Janina M. Jeff is the first African American to graduate with a PhD in Human Genetics from Vanderbilt University and graduate of Spelman College (class of 2007). She is currently a Staff Bioinformatics Scientist at Illumina, where she develops pipelines for content annotation, selection, and design of population genome-wide content as well as selection of clinically annotated variants for Illumina's genotyping array portfolio that enables healthy population screening. In 2018, Janina was selected as one of three winners (out of 18,000) from Spotify's Sound-Up Bootcamp for her podcast, "In Those Genes", is an international award-winning podcast that uses genetics to decode the lost histories of African descended Americans through the lens of Black culture. "In Those Genes" has been recognized by IndieWire, The New York Times, and Third Coast Audio Festival. Janina was recently named as one of the top 100 Influential African Americans by The Root magazine as well as Top 40 Under 40 Alumna with Spelman College, The National Quality Minority Forum and The Network Journal. Her TEDx talk, "Afrofuturism Through the Power of the Genome" , similar to her work, challenges the misuse of genetics information and empowers the Black community to learn the value that lies within their genomes work.

STEM PROs Questions

Why do you do what you do? / Do you have a defining/ah-hah/ eureka moment where you knew what you wanted to do?

I do genetics because to me it is the closest discipline of science to the community. Genetics is a rare science discipline that is mentioned at nearly every family event or gathering and in my opinion it is the most relatable. More specifically my eureka moment happened when genetics was the common answer as to why so many health disparities existed between human populations. This spurred my career in human genetics and then further into science communications by making genetics education culturally attuned and relatable to the Black community.

What is something you wish you knew while in school?

That you can't have success without failure. The first time I failed something I was devastated but it was ironically a turning point in my journey.

What is something you wish you did differently when you first started working?

I wish I saved more and paid more attention to the benefits offered at my company.

If someone wanted to do what you do, what's the best piece of advice you'd share with them?

Network, network, network!

If you could definitively answer one unanswered science question, what would it be and why? / What unanswered scientific question keeps you up at night?

What about disease pathology is truly a result of population genetics and what can actually be explained by social determinants of health?

What scientific discovery or event in history would you like to go back in time and witness?

This discovery of DNA!

What's a misconception about your line of work?

That certain people have one gene while other people don't have that gene. We actually all have the same genes but just slightly different versions of them.

What's the best part of doing what you do?

Meeting scientists from around the world and working from home.

What is a necessary evil in your industry?

Capitalism, lol!

Have you ever changed your line of work? If so, why and what was the change?

Yes, well slightly when I started my podcast, "In Those Genes" this was my entry into the world of science communications.

How has social media benefited or hindered your career?

Social media has provided me with a platform and tons of speaking opportunities. Social media gives a voice to those commonly silenced.

STEM PRO
EARYN MCGEE, PHD

DR. EARYN MCGEE: PhD & MS in Natural Resources & Conservation, BS Biology, Herpetologist and Science Communicator, Coordinator of Conservation Engagement at the LA Zoo

All platforms: *@afro_herper*
Website: *EarynMcGee.com*

DR. EARYN MCGEE is a Herpetologist and Science Communicator. Her passion for the natural world and all the people who live in it drives her work. She is most known for her popular social media game #FindThatLizard. Every Wednesday at 5pm MST, she post a photo of a lizard camouflaged in its natural environment and participants have to find it. The captions that go with each photo often give the players natural history facts about the lizards which double as hints on where to look. But Dr. McGee also uses this as an opportunity to talk about conservation and social issues. Keep up with Dr. McGee at @afro_herper on all platforms.

STEM PROs Questions

Why do you do what you do? / Do you have a defining/ah-hah/ eureka moment where you knew what you wanted to do?

I always knew I wanted to work with animals but I wasn't sure how. For the longest time, I thought I'd be a vet even though I knew that didn't feel right. After my freshman year in college, I was able to participate in field research. Spending a summer hunting for lizards in the Chiricahua Mountains was my eureka moment.

What is something you wish you knew while in school?

I wish I fully appreciated how hard it was to work a 9-5 and pursue my passion projects.

What is something you wish you did differently when you first started working?

N/a, I've only been out of grad school for 3 years so I think I need a bit more experience to answer this question.

If someone wanted to do what you do, what's the best piece of advice you'd share with them?

Make friends and have fun. The work will always be there. Take every opportunity that comes your way and be vocal about what you want.

What scientific discovery or event in history would you like to go back in time and witness?

I would love to see the world pre—colonialism. To fully experience the ecology and beauty of the world that's been lost.

What's a misconception about your line of work?

That I get to work with the Zoo animals

What's the best part of doing what you do?

For my day job, getting to the zoo before the public and walking around. For my passion projects, working on cool projects with even cooler people.

What is a necessary evil in your industry?

I don't think any evil is necessary. It's an active choice. I instead believe, and encourage everyone to believe in a better world.

Have you ever changed your line of work? If so, why and what was the change?

Yes! One of the changes have been to encourage other herpetologists to adopt the term lasso or any synonym as opposed to using "noosing" when referring to a lizard capture technique. Not only is the change more inclusive, it better describes what we do, and when engaging with non-scientists, they aren't scared off by the terminology. Many times when using the term "noosing" they believe we've hurt the lizard which is isn't the case.

What purchase under $100 has improved your life: career or personal?

My favorite pair of Teevas

How has social media benefited or hindered your career?

Social media has been a great tool for sharing my passion, finding friends, and learning about different professional development opportunities.

STEM PRO

LORENA SORIANO

LORENA SORIANO: Founder & CEO of every POINT ONE, Keynote Speaker, Forbes Fellow

IG: *@allthingslorena*

LORENA SORIANO believes science & tech products should be representative of the population that uses them. At the time of this writing she was the founder and CEO of every POINT ONE, a Seattle based public benefit corporation helping emerging tech companies embed sustainability into their company's DNA.

Lorena is a community leader, Keynote Speaker, Forbes Fellow and passionately supports those changing the world. A believer in "following your dreams and passions", Lorena left corporate America to pursue her own childhood dream of becoming a scientist and doctor. While completing her B.S. in Biochemistry she experienced the same problem she had as a child—lack of representation. This inspired her to create every POINT ONE, where she works with companies to develop and implement integrated sustainability and social responsible strategies.

Lorena actively shares on her Instagram @allthingslorena the behind the scenes of navigating balance as a new mom, lessons learned from startup life, biases in science & tech, tips from participating in 2 Microsoft for Startup Accelerator partner programs, and completing her graduate studies in Corporate Sustainability at Harvard.

STEM PROs Questions

Why do you do what you do? / Do you have a defining/ah-hah/ eureka moment where you knew what you wanted to do?

I believe that we all have the power to change the world by pursuing our passions, but it may take some time to realize what those are. I iterate in my life based on the experiences I have, what I learn from them, and how they line up with my values. It's scary to make those changes but doing so has yielded my greatest victories, both professionally and personally. I started my career in banking, moved to tech sales, pursued my biochemistry degree, accidentally entered the corporate sustainability space and am currently completing a masters in Corporate Sustainability. A common theme was the prevalence of poor or nonexistent socially responsible business practices. While I was in banking I witnessed common practices taking advantage of underprivileged customers. In tech sales I noticed that products as designed completely ignored a large number of customer demographics. While completing my degree I noticed that I was one of only two Latinas in my STEM courses. I started pulling at the thread to understand what was the root cause of these similar-but-different symptoms in so many spaces which lead me to Corporate Sustainability. My passion is to change the world for the better, and sustainability is how I do it.

What is something you wish you knew while in school?

Two big things: 1) There is no direct, linear path through life, and 2) find a problem you'd want to wake up every day to try and solve and go learn about it. With number 1, problems may require a combination of disciplines, degrees, and experiences. There is no cheat code where "if you do this thing then this other thing always happens". You're gonna have to pivot. For number 2, it's similar to "if you do something you love you'll never work a day in your life" but—in my mind—makes it more practical. There are problems that need to be solved everywhere. Find the one you care about and go tackle it. Even if you end up with a solution, odds are you'll have discovered a host of other problems you're passionate about solving.

What is something you wish you did differently when you first started working?

Not playing it safe. Hindsight is 20/20, but I went the "business corporate" route after graduation. Despite my interest and passion for science, as a first generation Latina, earning a degree in STEM seemed like a dream for a different person and the social pressure to pursue the promised financial stability offered by a job in business was overwhelming. I worked hard and had a successful career that delivered on that promise, but realized I was sacrificing a dream that *would* make me happy for a life that other people were telling me *should* make me happy.

If someone wanted to do what you do, what's the best piece of advice you'd share with them?

While there is a lot that could be said in regards to the importance of corporate sustainability in the future of emerging science, tech, and engineering, the best piece of advice I would share is that it is crucial to know where you want to go, understand where you are, and figure out what your first two steps need to be. Getting started is the hardest part so even if those two steps go in the wrong direction, iterate, learn, and adjust to work towards your goal.

What scientific discovery or event in history would you like to go back in time and witness?

I would love to witness when marine biologist and nature writer Rachel Carson realized the environmental and social harm caused by pesticides. I would especially like to witness how she addressed the skeptics of her time.

What's a misconception about your line of work?

"Sustainability = Environmentalism". While environmental sustainability efforts are crucial and an important part of sustainability it is only one of the pillars. The extreme focus on the environmental part of sustainability allows organizations to brush the inadequacies of their social responsibility and corporate governance policies under the rug.

What's the best part of doing what you do?

I'm helping build and design an inclusive, ethical, and socially responsible future.

What is a necessary evil in your industry?

"Band-aid" solutions—The problems that need to be solved regarding sustainability are neither simple nor straightforward. Band-aid solutions provide an initial framework to bridge gaps stemming from shortcomings in an organization's operations. They are "evil" more as a representation of how they are commonly used in sustainability strategy implementation. What happens all too often is that once a band-aid solution is installed, companies become complacent because they have slowed the bleeding. This relative stability is then viewed as sufficient and tolerable while progress stops on the journey towards addressing the root cause with an integrated and robust solution.

STEM PRO
AARON SHEPARD

AARON SHEPARD: Software Engineer, NASA

IG: *@spacecadetshep*
TT: *@spacecadetshep*
X: *@spacecadetshep*

AARON SHEPARD is a roboticist, electrical engineering student, entrepreneur, and a science storyteller! HeI first fell in love with STEM while watching the astronauts blast off in the Space Shuttle. His dream is to go to space one day. Aaron was crushed when NASA announced the end of the shuttle program during his freshman year of high school. At the time he thought that the space age was coming to an end and that he needed to grow up and aspire towards a "real" job. After high school, Aaron went to college and then onto med school (because that's what smart people do right). He thought that the stability and prestige of being a doctor would make him happy. He was wrong...

After a rough year of med school Aaron made the hardest decision of his life: To leave medicine for good and start over again. Aaron had no idea what exactly he wanted to do. He was confused, depressed, lost, and broke AF.

Aaron decided to go back to school for engineering. He thought that starting over would be the "happily ever after" of his career journey. It turns out that getting an engineering degree while working full time and being a real adult is difficult. Over the last 4 years he's struggled financially, failed (and repeated) multiple classes, questioned his own competency, and thought about quitting many times. But Aaron refused to give up!

Despite all of Aaron's struggles , he's done amazing things. Since 2016 he has:

- Gotten paid to build cool space robots through the NASA South Carolina Space Grant
- Fulfilled his childhood dream of working for NASA as an internNASA Langley Research Center
- Joined the larger scicomm community (that's how he met Kevin) and even did a TEDx Talk
- Founded @cogitobrains , a company that makes mind controlled tech!
- Aaron uses IG to share the lessons he has learned throughout his journey, inspire creativity through technology, advocate for Black lives, and occasionally show off the dope things he's building!

STEM PROs Questions

Why do you do what you do? / Do you have a defining/ah-hah/ eureka moment where you knew what you wanted to do?

I've always been interested in space exploration. When I was a little boy, I used to sit in a giant cardboard box in my living room and pretend it was my "spaceship". My ah-hah moment happened while I was a medical student. Instead of paying attention in class, I decided to watch the live stream announcement of a comet landing. I remember thinking to myself "right now some nerds are getting paid to launch things into space, and I want to be one of those nerds". Needless to say, I decided to drop out of med school and pursue a career in engineering instead.

What is something you wish you knew while in school?

The first time (2008-2012)—I wish I knew how to take advantage of all of the opportunities that NASA had available for students. At the time I remember being really interested in spaceflight, but not knowing how to actually pursue a career in it, so instead I chose to study something else. It wasn't until I was in school for the second time (2016-2022) that I figured out how to navigate all of the amazing aerospace opportunities that are out there.

What is something you wish you did differently when you first started working?

To be honest, I wish I was a little more confident in myself and my abilities. For the longest time I had impostor syndrome which made me hesitant to speak up in meetings. I was very afraid to make mistakes because I didn't want to be perceived as less intelligent by my colleagues. Over time I did realize that mistakes are just part of the process and I became more confident and comfortable in my job.

If someone wanted to do what you do, what's the best piece of advice you'd share with them?

Learn to work in a team! It's not about being smart, it's about being a good team player! Space projects are massive, and to be honest you spend a large amount of your time in meetings. Effective communication is key.

If you could definitively answer one unanswered science question, what would it be and why? / What unanswered scientific question keeps you up at night?

I know this may sound cliche, but I really want to know if life exists on other planets/moons/comets. The universe is so big, and we're discovering a lot of new exoplanets. By sheer statistics, there has to be life somewhere else in the universe right?

What scientific discovery or event in history would you like to go back in time and witness?

I wish that I could have been around NASA during the Apollo program. The culture at NASA leading up to the moon landings was exciting, innovative, and (for better or worse) very fast paced. Don't get me wrong, we're still doing cool and exciting things in space right now, but I feel like the moon landings will alway be a special moment in human history. Also whenever I meet someone who worked on the Apollo program, I can always see a little twinkle of pride in their eyes any time they talk about it.

What's a misconception about your line of work?

That everyone here is some kind of a Nobel prize winning super genius. Surprisingly, a lot of people at NASA (or in aerospace in general) are regular people who just so happen to be super passionate about putting people and things really high in the sky.

What's the best part of doing what you do?

The best part about what I do , besides occasionally realizing that this thing I'm working on is going into friggin space, is getting to share what I do with others. Any time I mention the work that I do, I can literally see people's faces light up and then they proceed to either ask me questions about space things or share something they learned about space recently. Space exploration is one of those things that makes a lot of people revert back to curiosity they felt as children and I love being able to make people feel that way.

What is a necessary evil in your industry?

Bureaucracy, red tape, and overly complicated processes for simple things. But alas, that's working for the government.

Have you ever changed your line of work? If so, why and what was the change?

Yup! At one point I was literally in medical school. While I'm sure it would have been alright to be a fancy doctor, at the end of the day my heart will always belong to spaceflight

What purchase under $100 has improved your life: career or personal?

Any one of those Arduino, or Arduino clones intro kits. They're a great hands-on way to learn the basics of electronics, circuits, and computer programming. They're also really powerful, so once you understand the basics you can use them to build some pretty cool things (like mind controlled robots).

How has social media benefited or hindered your career?

Social media is a very powerful tool that has definitely benefited my career. It allows me to connect with people all over the world who share my same interests and passions. I actually met a lot of my aerospace friends (including Kevin) on social media long before I met them in person. It also gives me a platform that I can use to educate others and help them understand why science (and scientific thinking in general) is so important in our daily lives.

STEM PRO

SUPREET "SUE" KAUR

SUPREET "SUE" KAUR: Lead Strategist at NASA Ames Research Center, MSc Space Studies, international Space University BS Industrial and Systems Engineering, San Jose State University

- NASA FIRST (Foundations of Influence, Relationships, Success, and Teamwork) Fellow
- NASA's 21st Century Technology & Innovation Fellow
- Brooke Owens Fellowship (Class of 2019)

IG: *@sue_aerospace*

SUPREET "SUE" KAUR is currently a technology strategist at NASA Ames Research Center in California. She helps develop and integrate multifunctional innovative technologies with internal and external stakeholders.

Outside of the technical domain, Sue is an aerospace educator and communicator at NASA. Outside the agency she leads DEIA and mental health awareness efforts in the aerospace industry.

Sue loves to travel the world, and often does so for work, or for the sake of her undying curiosity and sense of adventure. If Sue wasn't an engineer, she would probably be an art historian by day (surrealist, impressionist, and post-impressionist) and a standup comic by night.

In her free time Sue goes out to art shows or space-tech events, gets together with her family and friends (who affectionately describe her as "sassy" and "spicy"), tinker with home renovation projects, work out, or curl up with a nice mind-bending sci-fi book.

STEM PROs Questions

Why do you do what you do? / Do you have a defining/ah-hah/ eureka moment where you knew what you wanted to do?

I've never had that one defining moment after which I said, "this is what I want to do." What I do and what I want to do changes over time, and it will continue to do so in the future. Understanding the "what" and "why" has been a slow gradual process because I'm not particularly attached to a title, a technology, or a routine.

Years of reflection have gone into answering questions like, "What brings me joy and satisfaction in life?" "What skills do I possess? Where can they be utilized?" "What are my values? And my principles?" "Am I a leader or a follower?" "What legacy do I want to leave behind? And how do I want to be remembered?"

As it turns out, what I am AM concerned with is the impact I have on the world in the grand scheme of thing. I want to go to sleep knowing my efforts are doing good in the world, presently and in the future. I want to wake up excited to tackle the "impossible" task. I want to work with people who see me as a human and a catalyst, not a number or a cog in the wheel. As a leader I want to foster a community in which people have a voice and autonomy.

There isn't one singular role which meets the requirements for what I want to do—there are many. But THIS is what feels worth pursuing at this moment.

What is something you wish you knew while in school?

That your loyalty should be to yourself, to your lifestyle, and to your happiness—not an external entity. I love being a part of the aerospace industry, but it has taken a lot of unpaid and underpaid labor to get to where I am. There have been countless times when I have chosen my passion or told myself "This sacrifice is for the greater good" over my physical, emotional, or financial wellbeing—and the consequences of ignoring those needs came back to haunt me in the end.

Yes, you must work hard and study hard. Yes, you must step out of your comfort zone and take some risks. Yes, you must learn from your elders and gain experience. But continuously forfeiting your livelihood for a passion is not sustainable. And you don't see this in other adjacent tech industries. This isn't the norm.

When I first decided to take a stand against labor exploitation, by saying, "I'm not a martyr, I need to eat and make a living. If you want me to work for you, pay me a living wage," I expected to be pushed out or kicked out of the industry. While some people did question "my commitment", there were many others who saw my value and advocated for me.

It's up to you to define your value and set those boundaries.

What is something you wish you did differently when you first started working?

Taking more time to educate myself about finances and health.

I knew how to take something theoretical and make it practical. I knew how to make the "impossible" possible. I knew how to write requirements, perform trade studies, validate an experiment, complete the lifecycle of a project, but I didn't know how to do my taxes. Or the difference between a 401K and Roth IRA. Or what percentage of my paycheck I should be putting away for an emergency fund. Or how often I need to get a regular health checkup. Which stretches I should be doing if I am hunched over a computer for hours on end. Or how much water I need to drink for every cup of coffee consumed. Or how to make quick nutritious meals.

There is so much to adulthood that was never covered in high school. I learned all of this (and so much more) reactively the hard way, not proactively.

If someone wanted to do what you do, what's the best piece of advice you'd share with them?

You can either live the life that is prescribed to you and make peace with that, or you can create a life you find fulfilling. I chose the latter.

The job I have today didn't exist. Many of the jobs I've had over my career never existed; you won't find them in a job search listing. They were created, either by myself or by companies who wanted me to work for them.

So if you know what you are good at and what you want to do, but don't see a career path towards that, then don't be afraid to create that career path. Be vocal and advocate for your interests to craft your career path.

What's a misconception about your line of work?

The biggest misconception is that I am in the thick of excitement 100% of the time. At any given moment I am doing some adrenaline-rushing thing or jet setting across the world. Or that I am easily taking one win after another, consistently achieving.

What you see on social media, in the news, or on NASA TV is the polished final product. And it is just the tip of the iceberg—it does not acknowledge the behind the scenes work which goes into making those wins a reality. A decade of education and work experience have gone into becoming who I am today. It has all been very intentional. There were no shortcuts, no connections, or luck involved in getting to this point.

Out of the dozen or so things I work on, maybe one will gain public recognition. Years of intellectual and technical work are required to take an idea from theory to something practical and demonstrable. Countless drafts and revisions go into publishing one article. Hours of rehearsing and filming result in a 1-minute clip featured on NASA TV.

Don't compare your life to someone's highlight reel.

What's the best part of doing what you do?

Public engagement and education. I love teaching and explaining complex ideas in easy to digest pieces. In my current role I am continuously learning and working on some far-out technologies and concepts alongside some very creative and intelligent people. No two days are the same, nor do I get bored. But the real satisfaction comes from helping others understand the value of the work we do at NASA and getting others excited about the global impacts.

What is a necessary evil in your industry?

Documentation. It is time consuming. Tedious. Bores me to tears, yet absolutely necessary and a critical part of my job.

Have you ever changed your line of work? If so, why and what was the change?

I've hopped around many industries—from fashion to biotech to (now) systems engineering. I tried on many different careers and jobs in different parts of the world to see what fostered my personality and lifestyle. Where can I make the most impact? Where can I grow, both professionally and personally? Which career path meets my adventurous side?

Trying on different career paths to find one that fits is completely normal! Changing your mind and pivoting is also a logical approach. Taking a break to reflect and find what brings you fulfillment is totally fine! What isn't logical is staying in something you have outgrown, or committing to something which no longer serves your needs.

What purchase under $100 has improved your life: career or personal?

A portable external battery pack with built-in cables. I travel A LOT and work on the go. I juggle multiple phones and devices to get my work done. My current battery pack is 10000 mAh and has come to my rescue more times than I can count. Whether it be last minute work travel, flight delays, a power outage, lack of accessible outlets—my battery pack has me covered.

How has social media benefited or hindered your career?

Social media is a double edge sword. On one hand, it has connected me to individuals around the world with similar passions and struggles, kept me up to date on aerospace events and trends, and allowed me to share my curiosity and potential opportunities with a niche community.

On the other hand, the internet can be a jungle. The content available can be overwhelming and (often) inaccurate so there is an investigative element to it. From a sociological perspective, the rules of engagement we have in real life aren't replicated online; there isn't as much accountability so people behave in ways online they wouldn't in person. I've compartmentalized my social media into "educational" and open to the public, and "personal" accessible to close friends and family for a sense of privacy, but those boundaries aren't always respected.

My two cents: (1) Use social media in a way that benefits YOU. Whether that's consuming content to decompress, staying in touch with long distance friends, learning about new educational or employment opportunities, getting involved in a cause, etc. You're not obligated to have an online presence, post constantly, or interact with anyone if you don't want to. (2) Protect your peace: engage with people who add a positive value to your life and follow accounts which build you up (instead of selling you solutions to flaws you don't have).

"Do the research. Ask questions. Find someone doing what you are interested in! Be curious!"

–Katherine Johnson, mathematician whose calculations were critical to the success of the first and subsequent U.S. crewed spaceflights

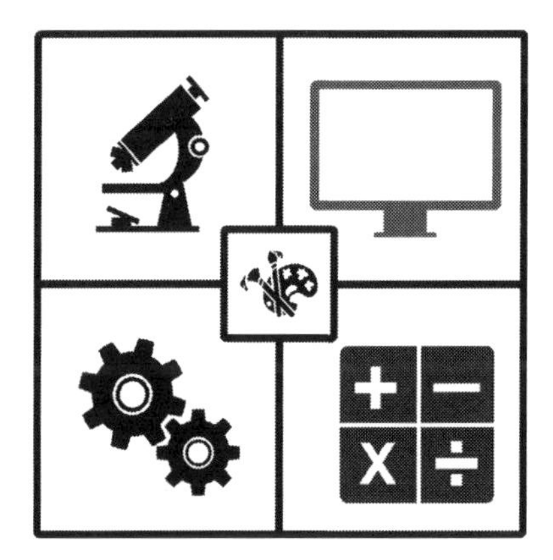

STEM PRO

MCKENZIE MARGARETHE

MCKENZIE MARGARETHE: A Queen marine scientist and naturalist, science communicator, LGBTQ+ Advocate, Post-Degree Diploma in Fisheries and Aquaculture, BS in forensic science minor in chemistry.

IG: *@_mckensea*
TT: *@mckensea*
X: *@missmckensea*
Threads: *@mckensea*
YT: *mckensea*
Website: *mckensea.com*

MCKENZIE MARGARETHE (any pronouns) is a marine science communicator and ocean conservation advocate. Currently they are in a post-degree diploma program for fisheries and aquaculture and have a passion for securing sustainable seafood systems during the climate crisis. Mckenzie also works diligently to uplift the LGBTQ+ community within STEM by educating others on the disparities the community faces and what can be done to mend the issues.

STEM PROs Questions

Why do you do what you do? / Do you have a defining/ah-hah/ eureka moment where you knew what you wanted to do?

I'm a marine scientist and at the time of writing this I'm currently receiving a post-degree diploma in Fisheries and Aquaculture from Vancouver Island University. My first love was the water. In fact, one of my favorite fun facts about myself is that I learned to swim before I learned to walk. However, I had been told repeatedly that marine science was underfunded and a career in this sector would lead to financial insecurity. This led me to originally pick a different science path, believing I'd keep ocean conservation as my hobby. After a year in my other program a professor came up to me to say I was doing incredibly well but it didn't seem like my heart was in it. She suggested I do a summer internship at the Hawai'i Institute for Marine Biology as they were looking for an explanary student to send to Dr. Ruth Gates lab. I did and my first day there I realized I was finally in my place and never looked back.

What is something you wish you knew while in school?

Probably to just start in a general biology/science program. I think it's unreasonable to expect 17/18-year-olds to know what career they want to get into as you never really know until you experience what it's truly like. Though some people know, I don't believe most people do fresh out of high school. I think I would have saved myself some serious time (and money) had I originally just gone into a general science program then done various internships to see where my passion is. The majority of STEM careers require an advanced degree (typically at least a masters) which means you can specialize at that point, or even add a concentration/minor during your undergrad. Give yourself the time to discover what fits best for you.

What is something you wish you did differently when you first started working?

To take advantage of opportunities when they present themselves! Many schools offer certifications like open-water SCUBA diving and scientific diving. Though they can be an additional fee they are typically much cheaper than taking it outside of school. If you're a student in Canada in the environmental sector there are many opportunities for early career professionals under the age of 30. I grew up in the states and therefore didn't know about these and only got to participate in one. I wish I could have done more of them to post my CV. I was also told by an advisor to not be afraid of rejection. Apply for so many opportunities that you're receiving acceptance/rejection emails constantly (and because you're applied for so many the rejections will roll right off your back).

If someone wanted to do what you do, what's the best piece of advice you'd share with them?

Give yourself time and grace. Marine science is a huge field, and you'll probably find many passions. Explore those passions, cast a wide net and experiment with different research areas. It's also okay to change your mind. People often say "giving up" on a field or changing out of it is a sign of weakness, but really it is a sign of strength. It is a sign you know yourself and what is going to fulfill your dreams.

If you could definitively answer one unanswered science question, what would it be and why? / What unanswered scientific question keeps you up at night?

The megalodon IS EXTINCT! Haha, no I'm just kidding—that's not unanswered. We as marine scientists know that definitively. However, I do often think of the extinct marine animals that we will never know about. We learn about the majority of extinct marine life from fossil records however only very particular conditions can create a fossil. Though we have discovered many incredible fossils (if you want a fun one look up the adventure to find the Tiktaalik), there are also countless marine animals that never created a fossil and therefore we will never know about!

What scientific discovery or event in history would you like to go back in time and witness?

As someone who has previously worked on a touring submarine and would like to work on a research submersible in the future, I think it would be incredible to go back and be one of the first people to reach the Challenger Deep—the deepest part of the Mariana Trench. The first time the Challenger Deep was visited was in the 1960s, which is an incredible feat. There were only two people on board and if I had a time machine (and with my current knowledge that they survived) I would become the third. Our continual discovery of the deep sea amazes me. Every year we are still learning about new species that live down there and so much of our ocean is currently inaccessible to us. Though we have used satellites to map our oceans and know their general geography I'm still excited for all the new discoveries that will come with truly exploring those untouched regions.

What's a misconception about your line of work?

That marine science isn't a "hard" science, that it's an easy field just looking at dolphins and sharks. This couldn't be further from the truth. Though we may be more lax in our appearance (I can't even begin to tell you how many marine science conferences I've attending with people wearing shorts and t-shirts), marine science is a serious science. You are required to know all the basics (many of us have taken organic chemistry, analytical chemistry, genetics, cellular biology, etc). Additionally, you must take that information and apply it to fish/marine invertebrate/marine mammal psychology and molecular processes. I wish my day was just spent on the ocean observing my favorite animals, but really most of my days are spent behind a computer studying and analyzing data. Did I mention you'll most likely have to learn how to code? Yea, add that to the list.

What's the best part of doing what you do?

It's what I love. At the end of the day, it's my true passion. As someone currently studying fisheries and aquaculture even many of the people in marine science look down on what we do. However, I truly believe in ocean conservation and global food security and that those two things needing to co-exist. Therefore, I am studying to find the solutions. Improving fisheries and aquaculture is the best way to feed our growing planet. Instead of arguing to dispel these fields I am trying to improve sustainability and public access to the facts. I believe this makes a lasting positive impact for not only our oceans but humanity. That belief is what I live for.

What is a necessary evil in your industry?

Ugh—unpaid work. Additionally, the cost of starting. Dive gear, lab gear and even getting experience can be EXPENSIVE. Marine science has a huge issue of "pay to play"—where you must pay to go on volunteer work experiences to initially build your CV. If you can't "pay to play" you often must settle for unpaid work in your region. It's a major issue in the field that people are actively working to change. However, as it stands the majority of beginner level work is unpaid or you have to pay for.

Have you ever changed your line of work? If so, why and what was the change?

Yes, I changed from initially wanting to get into forensics, as it's much more financially stable than marine science. However, it just wasn't my passion. Additionally, I've moved around in what field of marine science I want to commit to. In the beginning of my career I focused mainly on marine invertebrates, such as coral and urchins. I then switched to working with humpback whales and public education. Now I've incorporated a lot of that to studying the big picture or if you want to get technical—coastal marine ecology and species interactions with a focus on the impacts of the seafood industry.

What purchase under $100 has improved your life: career or personal?

A good field guide! Some of my personal favorites are "Coastal Fishes of the Pacific Northwest", "Marine Life of the Pacific Northwest", "Marine Fishes of Hawai'i" and "Whale of the Pacific Ocean". However, tailor this to your region and your passions. If you're currently in the area you'd like to study you can often find field guides at thrift/used book stores. This cuts your cost down and allows you to get several books. Field guides have taught me so much and are one of the best ways to familiarize yourself with different marine organisms and their abilities.

How has social media benefited or hindered your career?

A little bit of both. Some scientists really like to tease my social media presence and insinuate in makes me a less serious scientist. However, many others see the benefit of public education and commend me for the work I do. I find that it's the people who enjoy my content that are the people I want to be around and the others who don't aren't worth my energy or time.

STEM PRO

KARI BYRON

KARI BYRON: Mythbuster, Crash Test Girl, cofounder of EXPLR and National STEM Festival.

IG: *@therealkaribyron,*
@nationalstemchallenge,
@explr_media,
FB: *The Real Kari Byron*
X: *@karibyron*
TT: *@explrdotcom,*
@therealkaribyron
Websites: *karibyron.com,*
explr.com

KARI BYRON has been a strong presence in the world of science adventure and travel television for over two decades. She is best known as a host on Discovery Channel's Mythbusters and Crash Test World but has gone on to host and produce shows spanning several networks. Today you will find her continuing her mission to inspire, empower, entertain and educate through the digital storytelling of EXPLR and as a STEM advocate and founder of the National STEM Festival.

STEM PROs Questions

Why do you do what you do? / Do you have a defining/ah-hah/ eureka moment where you knew what you wanted to do?

I didn't come to my career in STEM in a linear way. I am an artist. I wanted to be in special effects. When I started my first day as an intern at M5 Industries to learn the trade, it was also the first day Mythbusters started filming. I went from cleaning up behind the scenes to building behind the scenes to telling a camera what I was building to being a presenter on a science based reality show. I didn't even know I was doing science. I was a maker. It wasn't until I approached science like I approached art, getting my hands dirty and quenching my curiosity with building an experiment, that I realized I love science. Art and science are all about asking questions and fostering a curiosity. It just so happens that the scientific method is also the best narrative vehicle for busting myths. Being a digital storyteller can absolutely be a STEM job. Science is a good story.

What is something you wish you knew while in school?

Take every opportunity to learn new things. Say yes. Being brave is about being scared and doing it anyway. I was bullied for being different when I was in middle school. I felt so alone, depressed and scared. It wasn't until I found my people, my community, the ones that were kind, empathetic and smart that I felt seen. I have met lots of kids who are STEM focused that have a hard time in their school community. I want to tell them, "it gets better". One big lesson that I learned from the National STEM Festival is that when all the student champions were in the room together, they found their community and those connections have become invaluable and have continued. It can be so hard to be different but it is also a super power. It makes you special and you can find other special people. Find them. STEM is stronger with collaboration and creating your STEM community will open so many doors; in your heart, in your world and for your future.

What is something you wish you did differently when you first started working?

I wish I knew my worth. As a woman in the early days of my industry, I was the first on camera science communicator on reality based tv and the only long running female presenter at the time. I was being paid way less than my male counterparts with the same job and title. I had to fight hard to be paid equitably. I hope I changed the scene for the next generation of women. I still advocate for those new to the field.

If someone wanted to do what you do, what's the best piece of advice you'd share with them?

The industry has changed entirely since I started. My best advice is to find someone doing what you want to do and ask for advice. Find mentors. Don't be afraid to work your way up.

If you could definitively answer one unanswered science question, what would it be and why? / What unanswered scientific question keeps you up at night?

Why do we sleep? I am obsessed with the concept of sleep. Since it makes us more vulnerable. Why would we evolve to have to sleep?

What scientific discovery or event in history would you like to go back in time and witness?

Tesla inventing his coil. Just madness.

What's a misconception about your line of work?

Storytelling isn't a STEM career. Busted. There are all kinds of STEM jobs.

What's the best part of doing what you do?

Mythbusters opened doors for me and now I can open doors for others. I became a founder of the National STEM Festival to elevate and celebrate this generation of leaders!

What is a necessary evil in your industry?

Social media.

Have you ever changed your line of work? If so, why and what was the change?

I have moved behind the camera. I love creating the stories but think sometimes the best person to be the storyteller isn't necessarily me. It is a collaborative creative process. I love every part of it.

What purchase under $100 has improved your life: career or personal?

Leatherman multitool

How has social media benefited or hindered your career?

I have a much thicker skin as it has become darker but also have been able to utilize it for good. It is a brilliant tool to meet kids where they are at to bring them where we want them to be. It is the new frontier of education and entertainment.

STEM PRO
CHELSEA CONNOR

CHELSEA CONNOR: Herpetologist, Grad Student, coFounder #BlackBirdersWeek, Researching Lesser Antillean Phylogeography for work and Anolis lizards for fun

X: *@ChelseaHerps*
IG: *@OutToChelsea*
TT: *@ChelseaBirds*
(I need to make them all the same thing at some point)

CHELSEA CONNOR is a herpetologist, artist, birder and PhD student from the Commonwealth of Dominica in the Caribbean. She has worked on the diet overlap between the native and invasive species of anole on her home island as an undergrad. As a graduate student, she is working on the phylogeography of Lesser Antillean reptile clades. Chelsea loves anoles and birds. As a science communicator she has spent time creating birding videos, and putting together and sharing information on different anoles with her hashtag #DidYouAnole? on Twitter.

STEM PROs Questions

Why do you do what you do? / Do you have a defining/ah-hah/ eureka moment where you knew what you wanted to do?

I grew up watching the Discovery Channel with my dad and I always wanted to be outside with animals. I have loved animals my whole life. But I never thought about working with animals the way I do now, until I met my undergrad advisor who told me about his work with anoles. He told me that having the opportunity to take part in the research in my home country would be great for me, and when I said I didn't know if I could do that, he just said "Why not?" And every time I've wanted to try something new I've been asking myself that ever since.

What is something you wish you knew while in school?

The many different pathways you could take to work with animals. I used to think that being a zookeeper or a vet was the only way you could get into an animal related job/field!

What is something you wish you did differently when you first started working?

Taking more chances. The worst that can happen is that someone says not right now or it doesn't quite work out but you learn from all of that.

If someone wanted to do what you do, what's the best piece of advice you'd share with them?

Keep your passion! Sometimes the data analysis and reading all the reference papers gets tiring. Especially when you're trying to complete something that should be simple or find one line of text that will make your paper. But you're doing it because you love reptiles, because you love learning and teaching, because you love seeing how it all ties together in the end.

If you could definitively answer one unanswered science question, what would it be and why? / What unanswered scientific question keeps you up at night?

My unanswered science question is probably silly. I want to know if there's flying anoles hidden in a forest somewhere and what forest would that be if they existed. Sometimes I lie awake thinking about what it could look like, what I would name it. It's already the most perfect lizard in the world to me and may never exist haha.

What scientific discovery or event in history would you like to go back in time and witness?

I read this paper from the 70s once that described the first instance of seeing this species of anoles jump into the water to escape when the researchers/threats approached. They then caught some and tossed them into the middle of the stream to see if they really could swim. I would like to have seen that. I feel so bad for the anoles, but walking up on them initially to see them leap into the water and swim away must have been mind blowing.

What's a misconception about your line of work?

That I handle lizards all the time. My current project doesn't even need fieldwork! I am relying on people who have done fieldwork already and made their data public.

What's the best part of doing what you do?

I really enjoy learning new things and teaching other people about it. Either through my actual professional research or personal. But I also love making niche animal jokes.

What is a necessary evil in your industry?

Animal specimens. I know that some people have issues with that, but it does serve a purpose and we have guidelines for every project that will involve an animal to make sure that only the necessary data gets collected from them, they're treated properly and they are safe if they are released.

Have you ever changed your line of work? If so, why and what was the change?

Yes, I saw The Mummy and decided I wanted to be an archaeologist. Then I wanted to be a vet in high school, but then I learnt how rough vet school can be and decided to be a zookeeper. But here I am now!

What purchase under $100 has improved your life: career or personal?

Procreate! It's an iPad drawing app and it's where I create all of my art. I didn't realise how terrible my old drawing program was until I got it. Now I'm able to create all the funny bird and lizard art I see in my head. Best $20 I've ever spent.

How has social media benefited or hindered your career?

Social media has definitely given me a boost and put me in touch with people I never would have gotten the chance to (Kevin included)! I've been able to share my passion and people who see that and like it get to engage with the things I put out and create. It's been great to see people light up when they learn something new or they enjoy my artwork. They get to share that with other people who come back to me on their own and reach out with opportunities or avenues to the things I've wanted to try. I don't know if I would quite be here without social media.

STEM PRO
NATALIA REAGAN

NATALIA REAGAN: Anthropologist, Primatologist, Comedian, Host, Podcaster, Producer, Writer, & Bonafide Weirdo. MA & BA in Anthropology/ Primatology

IG: *@natalia13reagan*
TT: *@beholdnatalia*
X: *@natalia13reagan*
FB: *Natalia Reagan Official*
YT: *@NataliaReagan*
StarTalk: *startalkmedia.com/bio/natalia-reagan*
Ologies: *alieward.com/ologies/gluteology*

NATALIA REAGAN is an anthropologist, primatologist, comedian, host, producer, podcaster, professor, writer, frequent animal expert, and monkey chasing weirdo. Sure, she wears a lot of hats, but her favorite one is her beloved Waldorf and Statler trucker cap. Natalia's science communication style was recently described as "hilarious, unhinged, and strangely hot". She'll take it.

As a child, Natalia had recurring King Kong nightmares where he ate her entire family, one by one. In her first lucid dream, she successfully shrunk Kong down to an orangutan. It gave her a hug. Since that moment Natalia was obsessed with primates. And all animals in general. She has been "beholding" all critters since she was knee high to a grasshopper. Steve (the grasshopper) is now dead, but she went on to study a critically endangered subspecies of spider monkey in rural Panama and now is an animal expert on multiple TV shows.

Before she became a scientist, Natalia was an actress and improv comic. She scored her SAG card playing a dancing Chicken McNugget in a McDonald's commercial. She went onto to appear in TV shows like "My Name is Earl", Everybody Loves Raymond", and played a pregnant prisoner with a black eye (type casting, amirite?) in the teen road comedy "Sex Drive". After getting hit by her own truck on the shoulder of the 101/405 interchange—she survived, the truck did not—she realized she wanted to focus on her first love, science. After an extensive recovery period, she went back to school to study biological anthropology—including her beloved gorillas. For her master's fieldwork, she conducted a survey of the Azuero spider monkey in rural Panama. She has also published chapters in the Wiley Encyclopedia of Primatology, including "The Copulatory Postures of Nonhuman Primates" (read: monkey sex positions), ACS's Hollywood Chemistry, and Congreso de Antropología Panameña.

Natalia was trained with a four-field approach to anthropology and she often examines topics through a bio-cultural lens. She then deftly finds ways to make said topics endlessly entertaining and laugh-out-loud funny. After grad school, Natalia began pitching science comedy TV shows. She believed science and comedy were the perfect fit, but execs in 2010 did not. So she began producing science comedy videos in her garage, covering such titillating topics as the evolution of boobs, butts, balls, and Bigfoot. The boobs video landed her a guest spot on the TODAY show talking tata with Ann Curry and things snowballed from there.

Natalia has been a comedy writer and correspondent on Neil deGrasse Tyson's StarTalk, regular host of the StarTalk All-Stars podcast, a science correspondent on Thrillist's Daily Hit, a science expert on History's UnXplained, skeptic on Travel Channel's Paranormal Caught on Camera, and she was the co-host on Spike TV's 10 Million Dollar Bigfoot Bounty. Natalia was also a writer and host for Discovery's DNews, Seeker, and TestTube as well as an animal expert on Nat Geo Wild's Everything You Didn't Know about Animals. She currently podcasts for Scientific American.

Aside from TV appearances and digital content, Natalia is an experienced live host, stand-up comedian, and has written and produced live science comedy shows at NYC's intellectual hotspot Caveat, including "That Human

Show!" and "Science For Social Change". Up next, she's producing a science comedy special called "Survival of The Filthiest".

In 2019, Natalia started making TikTok and IG videos as her "Behold Natalia" persona—think naughty David Attenborough meets Madeline Kahn—and has a produced hundreds of hilarious and informative videos about animal behavior, sexuality, and *ahem* junk. She also makes art with her boobs and has a greeting card line called "Smoking Chicken Cards". The tagline? "I draw animals that drink, smoke, and sleep around so I don't have to" (that hasn't stopped her).

Natalia believes comedy can help democratize science. It is the perfect method to reach the public and hold their attention! So, needless to say, her passion includes combining science and comedy to spread science literacy while inducing spit takes. She currently lives in LA and produces science comedy content while her hyper-critical pug, Raisin, quietly judges from her throne of lies (the couch).

STEM PROs Questions

Why do you do what you do? / Do you have a defining/ah-hah/ eureka moment where you knew what you wanted to do?

I want to democratize science using humor and heart! I love understanding the how, what, where, when, and why humans and animals do what we do. AND I want to share it with everyone who is willing to listen. That's why using comedy is such an obvious choice for me. Humor is an excellent way to engage audiences who might not be in the mood or 'into' science. If you get them laughing, you KNOW they're engaged! And that opens the door to let the accidental and intentional learning begin!

What is something you wish you knew while in school?

Life is not linear. It's a brambling bush of twists, turns, hard and happy accidents, that leaves you with so many invaluable lessons. As a professor, I often told my students "It is ABSOLUTELY ok and TOTALLY normal to not know what you want to do for the next 40 years. Plus, it's also ok to change career course if you want!" I didn't even know what anthropology was until I took my first class in college. College should be a time to explore new avenues and try new things! And honestly—I think life should always be about getting out of your comfort zone and diving into the deep end. That's where the exciting, innovative, and invigorating work happens.

What is something you wish you did differently when you first started working?

Work begats work. Even though you might get offered jobs that are not exactly what you want, they made open the door to other better opportunities. I've had a myriad of jobs throughout the last 27 years—from smoothie slinger at Jamba Juice, to commercial set dresser, to TV stand-in, to sitcom actress, to night tech at a rehab for adolescents, to Fosters beer girl, to biological anthropology professor, to primatologist in Panama, to TV comedy writer, to department store make-up artist, to dancing chicken McNugget, to Sony Studio shop girl/window dresser, to science podcaster, to Bigfoot TV show host, to Griffith Observatory museum guide, to Nat Geo Wild animal expert, to SyFy technical advisor and SO MANY MORE.. All these jobs have somehow opened the door to other opportunities OR were the source of invaluable life experience that I pull from on the regular basis.

Diversify! If you're only focusing on one platform, that means you aren't connecting with different audiences. That's not to say you should spread yourself too thin, but make sure you engage folks on multiple platforms—even just a little bit!

Also, as a science communicator, you will probably be asked to work for free A LOT. Heck, I still do! While working for "exposure" may appear necessary to get your foot in the door, do NOT fall for that exploitative tactic. By working for free, it hurts everyone in our field. And it's totally unfair to you and sets up the expectation that you will continue to work for free in the future. It devalues you, your experience, and education. Plus, if this employer who wants you to work for exposure found you, then you don't need the exposure! Since you clearly were easy to find. Essentially: Value your work and what you bring to the table. (FYI, doing podcasts and promotional appearances is different!)

If someone wanted to do what you do, what's the best piece of advice you'd share with them?

Learn the ins and outs of the fields you want to work in. If you want to become a one-stop shop science communicator, you will need to learn how to perform, write, shoot, edit, and adapt to constant change. You want to work in TV? Learn all about unscripted, nonfiction, scripted TV production work—top to bottom. You want to podcast? Learn how to record, edit, and produce a podcast. Or have the funds ready to pay someone to help you.

Oh, and speaking of funds—and this is a big one—COMPENSATE YOUR PEOPLE! Sure, some folks may insist on helping for free, and in some cases that's ok. But if you ask someone to do a special skill (photography, editing, shooting, make-up, etc) make sure you are prepared to offer them compensation. Like anything, you get what you pay for. And in some cases, you can exchange services with somebody. If you're great at Photography, perhaps you can get somebody to edit your podcast for free and exchange for new head shots. But at the end of the day, you want people to feel valued. That is when they do their best work.

If you could definitively answer one unanswered science question, what would it be and why? / What unanswered scientific question keeps you up at night?

I wonder what other animals are thinking.

I wonder why some humans cling to conspicuous consumption and greed.

I wonder how other great apes feel in captivity versus being in the wild.

I wonder about the "intelligence" of humans. Sure we developed Penicillin, built the atomic bomb, and painted the Sistine chapel, but can we unfuck ourselves out of anthropogenic climate change? I'm optimistic, but the greed of humans is our biggest downfall.

I think it's geo/species centric to assume we are alone in the universe, so I wonder who or what else is out there.

What scientific discovery or event in history would you like to go back in time and witness?

I would love to be a fly on the wall during the middle/late Pleistocene in Europe and see how anatomically modern humans (AMH), Neanderthals, Denisovans, and other extinct members of the genus Homo interacted. Thanks to ancient DNA, we know that AMHs, Neanderthals, and Denisovans all got funky with each other, but how did it play out? Did they bicker? Was it love at first sight? Was it forced copulation? I want all the dirty deets!

I also would love to see our ancestors, taking their first steps on two legs. How did that look?

And lastly, I would love to have been on the Yucatán Peninsula 66 million years ago when that asteroid (Chicxulub) struck the Earth and wiped out the dinosaurs and eventually gave rise to the mammals.

What's a misconception about your line of work?

That it is easy. That it pays well. That it is a simple alternative to teaching. Science communication is a combination of many skills—many of which take years to master. That's not to say it can't be done, but I meet a lot of folks that just want to pivot into science communication and are surprised that they aren't wildly successful within a few years.

What's the best part of doing what you do?

Educating folks while making them laugh their keisters off! Whether it's my stand-up act, my online content, or TV work, nothing makes me happier than folks coming up to me or commenting online that they not only laughed, but learned a great deal from my work! In some cases, it has changed their perspective on social issues, which is even more meaningful. When I get to dispel myths about race, sexuality, and sex and gender and people actually listen & learn, that makes me feel like I did a job well done.

What is a necessary evil in your industry?

Self-promotion and social media. Self-promotion is necessary to book work and connect to your audience. And social media is an excellent—albeit often frustrating—way to connect and stay engaged with the public. But social media can (and most likely will) open yourself up to trolls, 'well, actually' dinguses, misogynists, bigots, and all-around asshats. It can be painful and challenging. Remember to be kind to yourself and step away from social media when it feels too much. Your mental health is worth so much more than 'followers' and 'likes'!

Have you ever changed your line of work? If so, why and what was the change?

Yes. I became a scientist in my late 20s. I started making science comedy content in my early 30s. It's never too late to try something new!

What purchase under $100 has improved your life: career or personal?

"A Primate's Memoir" by Robert Sapolosky. It has changed my perspective on so many aspects of life.

An iphone/phone lavalier—good audio is CRITICAL for creating successful video content.

How has social media benefited or hindered your career?

Both! I've been making content for 15 years and I know now that you cannot take the comments personally. I started off as an actress 27 years ago, so I have had my body picked apart my entire adult life. This means I have a very thick skin. Think pachyderm thick. But I'm not immune to being affected. Sometimes it's hard when people constantly attack the way you look, think, or simply exist. And honestly, I know so many many others have been attacked far worse than I have. Especially women of color and those in marginalized communities. The Internet can be a wonderful world, but it also can be filled with nightmares and horrors. For this reason, remember that the cruel things that people post are a reflection of themselves, not you.

How has social media helped? One of the first videos I made was called "The story of boobs: The breast tail ever told". After pitching science comedy shows and hearing no too many times from production companies, I decided to take matters in my own hands and make my own content. I posted it on YouTube in spring 2011. It didn't become viral, but in June 2011, the TODAY show reached out to me because they were going to be doing a segment on America's obsession With sweater cows and they wanted me to come on and talk about it from an anthropological perspective. disappearance led to Stephen Colbert, making fun of us on the Colbert report and called for butt week. So I naturally did a rebuttal and mini videos about the evolution of the human booty. This led to even more videos and ultimately led me getting cast as the cohost on $10 million Bigfoot bounty. The producer first reached out to me on Facebook, an example of how social media can be helpful. And from there I started making my own shoulder content called talking shit with Dr. Todd and Natalia. That series led me to host/write for Nat Geo Wild & for Neil deGrasse Tyson on StarTalk.

STEM PRO

MARIANNE PAGUIA GONZALEZ

MARIANNE PAGUIA NGUYEN: Space Engineer

IG: *@marsexplorer_*

MARIANNE PAGUIA NGUYEN is an engineer at Blue Origin. She is also a former technologist and engineer at NASA Jet Propulsion Laboratory. She hopes to inspire the next generation of female engineers. She has worked on the Mars Perseverance rover, several instruments for the International Space Station, and Europa Clipper. Her latest project is Blue Alchemist, which aims to create solar cells directly on the moon's surface out of lunar regolith. She is passionate about space exploration and values developing technology in space to benefit civilization on Earth.

STEM PROs Questions

Why do you do what you do? / Do you have a defining/ah-hah/ eureka moment where you knew what you wanted to do?

As a kid, I grew up watching Star Wars. This sparked my curiosity about the universe. I firmly believe that art drives innovation, and vice versa. I never thought it would be in cards for me to be able to work in the space industry, but it was something I always dreamed of.

What is something you wish you knew while in school?

I wish I had learned that it was normal to face rejection and occasionally fail. I put a lot of pressure on myself which helped me succeed, but at times it was to a fault. It is normal to feel lost at times, that is the whole point of being a student—to learn and grow.

What is something you wish you did differently when you first started working?

I'm very passionate about my work, so when I started my career, I was prone to being burnt out. In hindsight, I wish I had given myself limitations and had established a healthy work life balance early on.

If someone wanted to do what you do, what's the best piece of advice you'd share with them?

Plenty of my peers in engineering school would do nothing but go to class. Don't be like that. Get involved with student organizations. Compete in STEM competitions. Join a research group. Start your own personal engineering projects and have fun with it!

If you could definitively answer one unanswered science question, what would it be and why? / What unanswered scientific question keeps you up at night?

The thought that most often keeps me up at night are the existential questions of "Are we alone in the universe? Are there higher dimensional or more advanced species than those on Earth?". If we find the answer to this question is yes, I am hopeful that more people will unify in their passion for science and space exploration.

What scientific discovery or event in history would you like to go back in time and witness?

I would really love to witness the first human flight. The invention of the airplane was the starting point that got us to outer space.

What's a misconception about your line of work?

Many people assume you have to be a genius to work in the space industry. At the minimum, you need dedication, integrity, and a willingness to learn from others.

What's the best part of doing what you do?

It is the most rewarding feeling to know that there are technologies operating space that I've had the privilege of contributing to.

What is a necessary evil in your industry?

Space is hard, and safety is a big component to that. Safety at times can seem like it gets in the way of innovation, but it is absolutely necessary. To be successful, we need to research and develop our work in careful steps.

Have you ever changed your line of work? If so, why and what was the change?

I never had a major change during my career in regards to discipline, but I have taken a huge chance and accepted a big engineering role for a project that was entirely out of my technical comfort zone. I was hesitant, but I trusted my peers and mentors who recognized that I would adapt well to and eventually excel at the role.

What purchase under $100 has improved your life: career or personal?

The first snowboard I ever bought was one of the best decisions of my life. Snowboarding fulfills my need for adventure and helps me appreciate how beautiful our planet is.

How has social media benefited or hindered your career?

Social media has sometimes made me vulnerable to haters who somehow find the energy to tear women in STEM down. But at the same time, social media has gifted me with the most supportive friendships and networks that I could have never imagined!

ABOUT THE AUTHOR

Kevin J DeBruin

Who am I? Well thank you for getting this far in the book and coming to read about me now! I saw the movie October Sky at 10-years old and knew I wanted to design spaceships for NASA. It took me 15-years to make that dream a reality while battling numerous bouts of adversity (people laughing at my dream as a child and a college student, NASA rejecting 150+ internship applications, Georgia Tech denying me admission into grad school, NASA JPL interviewing me but not giving me a job).

I left NASA because I fell in love with talking to people about space rather than sitting in a cubicle by myself designing spacecraft. Basically I was at my dream job and not happy. My introduction to space education was my 5th grade teacher asking me to come speak about my first NASA internship to her class during winter break from college. This experience gave me the most surreal feeling in the world, but I was still on my own journey and didn't even consider it being something I could do full-time.

I do what I do now because it makes me happy and it utilizes my full skill set where I am not expendable. At NASA, as soon as I left I was replaced. But now if I don't do something, it doesn't get done. I discover and invent my own ways to spread the good word of science to the public and make my own unique impact on the world and its inhabitants. I love having control of my time and creativity to teach about space in the way that I think it should be done.

KEVIN J DEBRUIN: NASA Rocket Scientist turned Entrepreneur, Author, Host, Speaker, MS Aerospace Engineering, BS Mechanical Engineering, Formerly NASA JPL, Formerly The Aerospace Corporation

IG: *@kevinjdebruin*
X: *@kevinjdebruin*
TT: *@kevinjdebruin*
Threads: *@kevinjdebruin*
YT: *A Place Called Space*
FB: *Kevin J DeBruin NASA*
LI: *@kevindebruin2589*
Websites: *kevinjdebruin.com, VirtualSpaceClass.com, ToDareMightyThings.com*

KEVIN J DEBRUIN is a former NASA Rocket Scientist turned Entrepreneur spreading space and motivation around the globe. In 2018 he left NASA and started his own Space Education & Consulting Company as an author, speaker, host, consultant, and public space expert. He is the founder of Space Class (VirtualSpaceClass.com), creator & host of A Place Called Space on YouTube, CuriosityStream's Brand Ambassador for all things Space & Science, a Host on The King of Random (TKOR), a 2x TEDx Speaker, international Space & Science Camp Instructor, and author of To NASA & Beyond and To Dare Mighty Things.

Kevin worked as a Systems Engineer at NASA's Jet Propulsion Lab (JPL) in Pasadena, CA. His focus was on the Europa Lander Mission Concept as a Flight Systems Engineer as well as more than 30 advanced mission concepts exploring the universe as a member of TeamX, TeamXc, and A-Team. Kevin is an expert in Model-Based Systems Engineering (MBSE) and offers a training course Fundamentals of MBSE with OpenX Education. After NASA he worked for a short while at The Aerospace Corporation as a Senior Systems Engineer where he launched two 1.5U CubeSats into Earth orbit.

Kevin fell in love with science communication while working at NASA JPL designing spacecraft. He became the most active member of the JPL

Speakers Bureau giving multiple presentations a week and even taking vacation to speak more frequently about space. He decided to leave NASA and dedicate himself to educating and inspiring others in science. Kevin's vision is to inspire and educate as many people as possible about the wonders of space exploration and its importance to us here on Earth. He works endlessly to bring space down to Earth for the masses in a fun & exciting fashion. With more than 30 mission concepts under his belt, Kevin is the go-to guy for solar system exploration.

Kevin was born & raised in Kaukauna, WI. He obtained a Bachelor's in Mechanical Engineering and a minor in Business Administration from the University of Wisconsin-Platteville and a Masters in Aerospace Engineering from Georgia Tech where he was Graduate Research Assistant (GRA) in the Aerospace Systems Design Lab (ASDL) under Prof. Dimitri Mavris.

Kevin is also a 2x American Ninja Warrior, Certified Personal Trainer, former bodybuilder, and an Eagle Scout!

STEM PROS Questions

Why do you do what you do? / Do you have a defining/ah-hah/ eureka moment where you knew what you wanted to do?

Since I've made a big career pivot, I'll give both my defining moments. At 10-years old I saw the movie *October Sky* and I knew I wanted to work for NASA. I remember sitting cross legged on the carpet in front of the tv in the living room watching this movie in awe. Then it was in graduate school that I learned about Europa, a moon of Jupiter that has the most significant astrobiological potential (aka the best place to look for aliens) in our solar system and I became hooked on the science aspect of space. Up until this point I was all about the engineering, the machines that did the exploration, but when I discovered Europa and learned about the other incredible moons in the solar system I was shocked and started to be actually interested in the exploration.

I changed my career a few years after working at NASA to leave and start my own Space Education & Consulting company because I fell in love with educating about space and realized I didn't want to be a full-time engineer anymore. The ah-ha moment here is in two-parts. The first was reading

Lewis Howes's book *The School of Greatness*. He listed the habits of the richest individuals in the world. As I read through that list I thought, "Umm, I do every single one of these, no joke, but there is no way to increase my income within NASA. I could never be one of those rich people rewarded substantially for their efforts." Do I do side hustles to make more money? Yes, but that still doesn't change my reality of working forty–sixty hours a week in a cubicle, trading time for money.

The second story is the one that really hit my soul, hit it deep, like an impactor. (That's a spacecraft that we intentionally crash into the surface of something to expose subsurface material—we literally impact the surface to create a crater.) We threw a retirement party for my boss's boss. He gave a speech in which he said, "Now I can finally live my life the way I want. I can wake up when I want, I can go to bed when I want, I can go wherever I want." He looked like he was having the happiest day of his life. He's retiring... he's like sixty-five... yet he can't go outside without a hat on because of a condition.... Whoa. Dude, I'm twenty-five years old at this party, and you're telling me I have to wait forty years to feel that way? That I have to "pay my dues" to live my life the way I want when I don't even have much life or health left? OH, HELLS TO THE NO. Nope, this is not the life path for me.

What is something you wish you knew while in school?

I wish I would have been exposed to the science of outer space earlier. Yeah we talked about the planets in 3rd grade, but I didn't learn about the moons until graduate school when I was working with NASA JPL on the Europa Clipper. I remember googling "What is Europa NASA?" having no idea what it was. This brings me to my main point of wishing I would have taken science classes more seriously like chemistry and biology. I didn't think I'd need them that much with engineering, like I dropped Advanced Chemistry senior in high school to become a gym aide. It would have given me more motivation and meaning to learning the sciences and retaining that information.

What is something you wish you did differently when you first started working?

I don't think I really have anything here because I was pretty proactive. It was something I did in my early days working my first "big boy" job at NASA JPL. Each of the more senior colleagues I encountered that I felt

I vibed with, I would ask "If you could go back to when you first started working here (or the job you first had), what do you wish you did?" I got things like "leave your work phone at work", "don't put work email on your personal phone", "say 'no' more", "don't work on weekends, even in the beginning, because it'll set a precedent that's hard to escape."

If someone wanted to do what you do, what's the best piece of advice you'd share with them?

For this one, well it depends on what part of my life you wanted to do. The NASA Rocket Scientist, the Public Speaking & Content Creation, or starting your own company? I don't really like the "follow what you love and the money will follow" because it's BS. You gotta work for it, you gotta work hard. Find what you love and then get really really good at it AND figure out how to generate revenue from it. So my best piece of advice would be "always be your true genuine authentic self". It's what landed me my job at NASA, connecting with people who could feel my passion; I really wanted it and I didn't stop until I got it. It's also what led me away from NASA and starting my own thing; looking inward to see what I wanted rather than what others/society/my parents wanted me to do. And it's kept my mindset and brand intact so far, turning down gigs/opportunities that don't align with who I am just to make money or get attention/gain followers.

If you could definitively answer one unanswered science question, what would it be and why? / What unanswered scientific question keeps you up at night?

Are we alone? But not really that question. I want to know if Biology is universal. If Biology is on other places in our universe. We've seen that Physics, Chemistry, and Geology work throughout the universe, but we have not yet seen biology other than Earth.

What scientific discovery or event in history would you like to go back in time and witness?

When life started on Earth. The moment of single-cell organisms and the moment life became complex.

What's a misconception about your line of work?

That I am a social media influencer. I am not. I am an educator. I don't make much money off of the socials, I rarely do promotions/sponsorship.

There are many types of science people in this world and on the internet. Let me give you a breakdown. One is not necessarily only in one of the following categories; most are a combination of two or more. Also one is not necessarily better than the other; it's all perspective and personal opinion.

There are 4 Categories:

1. Science Entertainer
2. Science Influencer
3. Science Educator
4. Science Person

Entertainer/Influencer *(I've combined these two together because they are quite similar, but I'm assuming you can decipher a difference)*

- To ignite admiration and jealousy
- Jumps on trends, hot topics
- Look at me, see where I am/what I'm doing
- Doesn't teach, shows "cool"ness and things
- Service to self or fellow entertainers/influencers—90/10 split
- Followers, top dog, possibly entitled
- Flaunts achievements
- Sponsorships often

Educator

- To educate and inspire, inform
- Stays in lane, topic of expertise
- Personal shares to create relatability and connection to audience
- Service to others—90/10 split
- Community, teacher, part of the family
- States credentials

Person

- Is a STEM individual
- Shares their work and opinions
- Participates in science discussions

What's the best part of doing what you do?

Seeing the inspiration and excitement in others. Also hearing stories later down the line of how my work has impacted someone. For example someone who heard me speak in school and now is majoring in engineering, or just got their job at NASA. Or found my youtube channel and is now enrolled at Harvard.

What is a necessary evil in your industry?

Social media, I honestly would love to never open a social media app again. Or needing to be "on", sometimes I really don't feel like being on camera or on-stage, but I have to "put on face" and do the work to make things happen. It can be exhausting. Negotiations too. I would love to do what I do for free and still be able to live life, but that's not how the world works. I need to charge people for my products and services in order to create more products and services to reach more people (and rent, food, utilities, etc.). This book is NOT one of those though, I've been distributing this book at cost and even invested a good chunk of my own money into the editing and publication process just because I think this information should be out there!

Have you ever changed your line of work? If so, why and what was the change?

Yup! I quit my dream job at NASA to start my own space education and consulting company.

So I got the dream job & then I quit. Why? Well I ultimately wasn't happy. I felt held back, controlled, and limited in my life. I yearned for more, I felt my potential halted. I got really into personal development when I read Lewis Howes's book The School of Greatness. In it he listed the habits of the richest individuals in the world. I thought, "Umm, I do every single one of these, no joke, but there is no way to increase my income within NASA. I could never be one of those rich people rewarded substantially for their efforts and have control of their time." About 6-months into working, my boss's boss retired and he gave a speech saying "This is the happiest day of my life. I can get up when I want, go on vacation when I want, do what I want." And the dude was 65+ years old and couldn't go in the sun without a hat on because his doctor said he had a condition. There was no way I was waiting 40 years to live my life the way I wanted to. So a combination of those two experiences, falling in love with educating about space, and realizing that there wasn't a prominent science figure for the public younger than Bill Nye, Neil deGrasse Tyson, & Stephen Hawking lit the fire to tell me I needed to be that person. So I set off to make that happen pitching myself to managers, agents, production companies, producers as "I'm Bill Nye meets The Rock". I created several show treatments, applied for auditions, and started making my own education content in 2016. I was one of the original space people on social media, now it's oversaturated with unqualified and non-credentialed space enthusiasts.

I used to be deathly afraid of public speaking until I spoke about Space. I was sharing my passion, my love, to others and it felt right & natural. I spoke about Space as frequently as possible in public and online. By the way, at NASA I couldn't make any money because I was representing NASA and I had to abide by the NASA redtape. BUT if I said I was an aerospace engineer in the industry, I could get paid & do it my way. I decided to leave NASA in 2019 and go to The Aerospace Corporation in El Segundo as a sole means of transitioning to being on my own. Because now that I wasn't at NASA, I could use NASA to promote myself. And boy did I do that, NASA everything! Haha. And it eventually worked, where 10-months later I quit the 9-5 after I landed a contract with Curiosity Stream as their Brand Ambassador for everything space and science.

What purchase under $100 has improved your life: career or personal?

I'm going to say my hammock because I am laying in it right now in my backyard answering these questions. It's also been one of my favorite spots to read or just relax. Everyone that comes over goes "oh, I should get a hammock" and then they do.

How has social media benefited or hindered your career?

It's done both. It's granted me access to the world of individuals and exposure for opportunities. But it's also hindered me in others doing stuff for free where I change, just because they want to do it or have a 6-figure salary paying their bills. Or that my follower count isn't as high as others, where some base metrics on the amount of reach possible rather than my credibility stamp of approval on something.

CLOSING THOUGHTS

May you have learned at least one thing new and broken at least one preconceived notion of what it means to be a STEM professional. Revisit this book throughout your academic and career journey. Each time you come back and flip through the pages something new will resonate with you. As you evolve professionally, the advice and experiences in this book will hit differently. Whether you're a student learning more about opportunities, a mid-career individual thinking about a change, or a tenure professional, share what you've taken away from this book with others so that they may also gain the STEM PROs wisdom from others.

"Nothing in life is to be feared, it is only to be understood."
–Marie Curie, physicist, chemist, and winner of two Nobel Prizes known for discovering radioactivity, radium, and polonium

"Everything is theoretically impossible, until it is done."
–Robert A. Heinlein, aeronautical engineer and science fiction author

"The best way to predict the future is to invent it."
–Alan Kay, computer scientist and winner of the A.M. Turing Award for his contributions to object-oriented programming languages and personal computing

"We have the opportunity to create the future and decide what that's like."
–Mae Jemison, former NASA astronaut, engineer, physician, and the first Black woman to travel into space

"However difficult life may seem, there is always something you can do and succeed at."

–Stephen Hawking, theoretical physicist, cosmologist, author, and recipient of the Presidential Medal of Freedom

"If you take a look at the most fantastic schemes that are considered impossible... you realize that they can be possible if we advance technology a little bit."

–Michio Kaku, theoretical physicist, co-founder of string field theory, and author

"Above all, don't fear difficult moments. The best comes from them."

–Rita Levi-Montalcini, neurologist and joint winner of the Nobel Prize in Physiology or Medicine for the discovery of nerve growth factor

"Most people say that it is the intellect which makes a great scientist. They are wrong: it is character."

–Albert Einstein, theoretical physicist known for developing the theory of relativity and winner of the Nobel Prize in Physics

"An investment in knowledge pays the best interest."

–Benjamin Franklin, American Founding Father, scientist, inventor, diplomat, statesman, and writer known for numerous inventions, including the lightning rod

"Do the research. Ask questions. Find someone doing what you are interested in! Be curious!"

–Katherine Johnson, mathematician whose calculations were critical to the success of the first and subsequent U.S. crewed spaceflights

Made in the USA
Las Vegas, NV
16 November 2024

3946b6be-83b2-4980-b714-87f6a5d752c3R02